AF556848

SOLVENT EXTRACTION AND ION EXCHANGE IN THE NUCLEAR FUEL CYCLE

SOLVENT EXTRACTION AND ION EXCHANGE IN THE NUCLEAR FUEL CYCLE

Editors:
D. H. LOGSDAIL AND A. L. MILLS
Chemical Technology Division,
Atomic Energy Research Establishment,
Harwell, Oxfordshire

Published for the
SOCIETY OF CHEMICAL INDUSTRY
London, by

ELLIS HORWOOD LIMITED
Publishers · Chichester

First published in 1985 by

ELLIS HORWOOD LIMITED
Market Cross House, Cooper Street, Chichester, West Sussex,
PO19 1EB, England

The publisher's colophon is reproduced from James Gillison's drawing of the ancient Market Cross, Chichester.

Distributors:

Australia, New Zealand, South-east Asia:
Jacaranda-Wiley Ltd., Jacaranda Press,
JOHN WILEY & SONS INC.,
G.P.O. Box 859, Birsbane, Queensland 40001, Australia

Canada:
JOHN WILEY & SONS CANADA LIMITED
22 Worcester Road, Rexdale, Ontario, Canada.

Europe, Africa:
JOHN WILEY & SONS LIMITED
Baffins Lane, Chichester, West Sussex, England.

North and South America and the rest of the world:
Halsted Press: a division of
JOHN WILEY & SONS
605 Third Avenue, New York, N.Y. 10016, U.S.A.

British Library Cataloguing in Publication Data.
Solvent extraction and ion exchange in the nuclear fuel cycle –
(Ellis Horwood series in industrial chemistry)
1. Nuclear fuels
I. Logsdail, D.H. II. Mills, A.L. III. Society of Chemical Industry
621.48'335 TK.9360

ISBN 0–85312–842–1(Ellis Horwood Limited)
ISBN 0–470–20241–6(Halsted Press)

Printed in Great Britain by The Camelot Press, Southampton

Contents

FOREWORD 7

The role of solvent extraction and ion exchange in the nuclear fuel cycle. 9
R.L. Nelson

SOLVENT EXTRACTION AND ION EXCHANGE APPLICATIONS IN ORE PROCESSING AND URANIUM ENRICHMENT

Present and future use of ion-exchange in uranium ore processing 17
D. Naden, J.P. Pilot and C.W. Renner

Solvent extraction in uranium ore processing 35
G.M. Ritcey

Uranium enrichment by solvent extraction 53
J.M. Lerat and C. Lorrain

FUEL REPROCESSING

Reprocessing of the fast breeder reactor fuels in France 65
L. Patarin

Fast reactor fuel reprocessing experience in Germany 73
K. Ebert

Fast reactor fuel reprocessing in India 85
G.L. Balasubramanian

Experience and trends in irradiated fuel reprocessing in Italy 93
A. Moccia

Status of the Breeder Fuel Cycle in the United States 101
W.D. Burch

UK experience in fast reactor fuel reprocessing 112
J. Frew and D.W. Harris

SOLVENT EXTRACTION AND ION EXCHANGE IN RADIOACTIVE WASTE TREATMENT

Survey of solvent extraction and ion exchange in radioactive waste processing 127
K.W. Carley-Macauly

Solvent extraction and recovery of the transuranic elements from waste solutions using the TRUEX process 137
E. Horwitz and W. Schulz

The application of inorganic ion exchangers to the treatment of medium active effluents 145
E.W. Hooper

The use of coordinating copolymers to extract ^{106}Ru from simulated nuclear wastes 157
M.J. Hudson and A. Dyer

ION EXCHANGE IN FUEL POND AND WATER TREATMENT

Ion exchange in CANDU experience 171
B.R. Nott

Ion exchange experience at the Winfrith SGHWR 181
G.C.W. Comley and D.J. Ferrett

Ion exchange developments in the management of CEGB fuel ponds and associated wastes 199
D. Bradbury, D. Broadbent, G.A. Fitchett, T.J. Martin and D. Reed

The role of ion exchange in the Three Mile Island clean-up 209
R.M. Wallace

INDEX 219

Foreword

In the space of less than half a century the nuclear industry has grown from splitting atoms of uranium to the world-wide production of energy. In the formative stages of the industry alternative separation processes were studied for the extraction and purification of uranium and thorium, the separation of uranium isotopes, the recovery of uranium and plutonium from irradiated fuel and the treatment of waste streams. As the industry matured the two most widely-applied separation processes in the nuclear fuel cycle were liquid-liquid extraction, more commonly known as solvent extraction, and ion-exchange. Though these processes are related in several respects each has found its own areas of application, the former for concentrated solutions during the reprocessing of irradiated fuel, and the latter for the more dilute waste solutions.

Due to the economic recession and intense debate of environmental issues during the last few years, growth in the nuclear industry has been slow, but with expectation of an upturn in the economy it seemed opportune to hold a meeting on the current status and future direction of the application of solvent extraction and ion-exchange in the nuclear fuel cycle. It should enable those engaged in these topics in many countries to gather together, share their experiences and discuss the current status of national programmes and recent developments. The meeting is a natural successor to one held at the Dounreay Nuclear Establishment in the United Kingdom during 1979 on fast reactor fuel reprocessing.

The meeting has been organised by the Solvent Extraction and Ion-Exchange Group of the Society of Chemical Industry in collaboration with the UKAEA's Atomic Energy Research Establishment, Harwell and the Process Technology Group of the Industrial Division of the Royal Society of Chemistry. The Solvent Extraction and Ion Extraction Group of the SCI organises several meetings each year, covering a wide range of subjects and is pleased to be involved in collaborating with Harwell and the Royal Society of Chemistry in arranging this meeting concerning the nuclear industry. Harwell has always had an interest in the application of solvent extraction and ion-exchange in the nuclear fuel cycle, from ore extraction through fuel reprocessing to waste treatment and those attending the meeting will have the opportunity of seeing some of the current work. The Process Technology Group of the Royal Society of Chemistry is concerned with the development of processes throughout the chemical industry and recognises the pioneering work of the nuclear industry in the fields of solvent extraction and ion exchange.

The organisers would like to record their appreciation for the support and encouragement of the UKAEA and the Director of Harwell, Dr. L.E.J. Roberts, without which the conference could not have taken place. We are also indebted to all the contributors to the proceedings who have provided the international flavour to the meeting.

Our thanks are extended to all the Harwell staff, too numerous to name, whose support and assistance have made the conference possible.

D.H. Logsdail
A.L. Mills

Solvent Extraction and Ion Exchange in the Nuclear Fuel Cycle

R.L. Nelson, Head, Chemical Technology Division, UKAEA, AERE Harwell

The application of solvent extraction in the chemistry of uranium is first recorded by Peligot in 1841 who noted the solubility of uranyl nitrate in diethyl ether and the ability of ether to extract uranyl nitrate from aqueous nitric acid.

For nearly 100 years since that time solvent extraction was mainly a tool of analytical chemists and it was not seriously applied on an industrial scale until World War II other than in a restricted manner in petrochemicals. The subsequent expansion of the nuclear industry due to the development of nuclear power programmes in many countries has led to a significant application of solvent extraction to nuclear fuel reprocessing, uranium metal preparation, uranium isotopic enrichment and to waste treatment from nuclear fuel reprocessing.

Ion exchange phenomena have been known for many years, particularly in the treatment of water for industrial and domestic purposes, but it was mainly due to Spedding and his co-workers that ion exchange processes began to resemble what we know today.

The similarities between solvent extraction and ion exchange processes and indeed their overlap with the so-called liquid ion exchangers are worth noting.

In general terms for either solvent extraction or ion exchange, the required species are usually in an aqueous solution and this is contacted with an immiscible phase (an organic liquid extractant or an ion exchange resin) which effectively removes the required value from the aqueous phase.

The extracted species can then be recovered into an aqueous phase if required and the extractant or resin recycled.

Unlike solvent extraction, ion exchange is generally used with a fixed exchanger bed whereas solvent extraction is largely a countercurrent flow process, although work on an industrial scale for countercurrent ion exchange processes is well developed.

Ion exchangers exhibit anion or cation absorption properties with both organic and inorganic exchangers being available. Anion/cation systems in solvent extraction are available with the amine extractants.

The design procedures for the design of solvent extraction and ion exchange flowsheets are similar, the choice of extractant or exchanger is wide and governed by the chemistry of the species being extracted, the ease with which it is extracted or adsorbed and stripped or eluted.

It is not surprising therefore, to find that in the nuclear fuel industry there are a number of processes where solvent extraction and ion exchange can be interchanged. The high specificity of ion exchange resins compared with many solvent extraction reagents, their relative radiation stabilities and to some extent their costs, tend to define where one might be used rather than the other in the nuclear fuel cycle.

The resins described by Spedding - cross-linked polystyrenes - have been the subject of much subsequent development and more recently the introduction of new types of inorganic exchangers for certain duties in the fuel cycle should be noted.

Similarly, the flammability hazards associated with diethyl ether as an extractant have led to the development of processes based on tri-n-butyl phosphate (TBP) and other extractants.

The present meeting is to review these applications in detail and to look into the future to forsee further developments.

IRRADIATED FUEL REPROCESSING

The reprocessing of nuclear fuel to recover pure fissile and fertile material is the centrepiece of the fuel cycle. The general processes are sufficiently well known that they will not be discussed in detail but some comparisons are perhaps appropriate.

Compared with most chemical and metallurgical processes the

recoveries from nuclear solvent extraction processes are exceedingly high, but they are fairly commonplace for the now widely used solvent extraction techniques used in extractive metallurgy. Not only are the recoveries high, but the products are also very pure.

Expanding the examination of fuel reprocessing to cover the transuranic elements, especially Am to Cf one sees a mixture of solvent extraction and ion exchange in use on a production scale. Moving beyond Cf to Lw, No and further, ion exchange remains the only practical technique at the present time for the separation of the heavy elements, thus enabling their chemistry to be confirmed on the basis of quite literally a few atoms.

URANIUM RECOVERY FROM ORE AND ENRICHMENT

All nuclear fuel is based on uranium or thorium and most nuclear power stations employ uranium fuel. Uranium ores are widely dispersed throughout the earth's crust with several major ore bodies occurring in different parts of the world.

Most of the uranium mined at the present time is in areas where electrical power is scarce, hence the on-site preparation of yellow cake and its transportation for further conversion. Since neither solvent extraction nor ion exchange are as energy demanding as conventional metallurgical techniques then the solvent extraction or ion exchange processes can sometimes be carried out at the mine itself.

Natural uranium contains about 0.7% of the fissile isotope ^{235}U. The first industrial-scale separation of ^{235}U used calutrons at the Oak Ridge National Laboratory in the USA in the 1940's. These machines were essentially manually operated and required a significant power input. Gaseous diffusion through membranes superseded the calultron, but these plants occupy large areas and again have high power requirements.

Gas centrifuge techniques are an obvious method for uranium isotope separation given the appropriate engineering technology. The technology was developed in the 1960s and led to the development of the gas centrifuge which has tended to displace gaseous diffusion.

The isotopic mass differences of ^{235}U and ^{238}U suggests that separation should be possible by solvent extraction since the diffusion rates for chemically identical compounds of ^{235}U and ^{238}U should lead to different extraction kinetics. Simple calculation will show that large numbers of stages are required to effect any reasonable degree of separation, but given the demand for low cost, low enrichment and the relatively high cost of energy, enrichment by solvent extraction is now becoming (technically and economically) feasible. Similarly redox chromatography using columns packed with ion-exchange resin provides another technique for uranium enrichment.

WASTE PROCESSES

A major concern of both the nuclear industry and its opponents is that of radioactive waste. Although it should not be suggested that the conversion of highly active waste to a non-leachable form is an easy technology, it is the so-called medium and low active arisings that had less attention until recently and it is these latter wastes that are of interest here.

So far, solvent extraction has not played a significant part in waste processing, although such processes such as TRAMEX for the removal of actinides from the highly active aqueous waste should be noted.

Ion exchange, by virtue of its high selectivity and its more favourable aspects for dealing with dilute solutions, is perhaps more generally appropriate to the waste field. Organic ion exchanger resins have the virtue that when loaded they can be incinerated to give low volume residues consisting mainly of, say, the oxides of adsorbed species, which are then available for storage or further treatment. Volume reduction is gaining in importance but incineration is not a simple technique.

The newer, inorganic, exchangers are also species specific but are more resistant to irradiation and can be used to retain their adsorbed species for total incorporation into a solid, for example, concrete.

Without wishing to raise the question of the ultimate disposal or storage of radioactive wastes, the ion exchange properties of soils, rock formations and clays should not be overlooked. The treatment and disposal of spent liquid extractants is another area which requires considerable attention.

REACTORS AND STORAGE PONDS

Current world interest in water-cooled reactors would not be possible without close control of their water chemistry. It is in the field of water chemistry that ion exchangers predominate and will almost certainly continue to do so as long as the need for high purity water remains.

Fuel storage pond water tends to pick up adventitious materials together with corrosion products and even fission products in the event of irradiated fuel being damaged whilst in store. The contaminated water not only poses an operator hazard but unless cleaned up its disposal is not permitted and prolonged use could enhance corrosion. Again ion exchange resins are best used in this type of clean-up.

INCIDENTS

Finally we come to that part of the fuel cycle none of us wish to see, the incident clean-up. The record of the nuclear industry in respect of safety is exemplary when compared with many other industries, but there is no room for complacency. It is important to ensure that when an event such as happened at the Three Mile Island plant does occur, it is essential that there is the technology available with which the clean-up operations can be carried out.

Decontamination is a routine operation in fuel reprocessing and handling, but following the decontamination process there is a need to decontaminate the cleaning solution prior to its disposal. Again the high specifity of ion exchange, its ability to deal with large volumes of dilute solutions, and the relatively simple equipment required for its application are of considerable value, provided that care has been taken in selecting the composition of the original decontaminant.

CONCLUSION

The Nuclear Fuel Cycle is now part of the everyday life of many countries in the world today and it would be surprising if many of the other countries with developing industrial bases did not aspire to the nuclear fuel cycle. Over the past 40 years it has become a highly sophisticated part of a nation's economy and both solvent extraction and ion exchange play key roles in the Fuel Cycle.

Prior to the 1940s, both solvent extraction and ion exchange were regarded as laboratory or small-scale techniques, with few large-scale industrial application. Today these processes have reached maturity, not only via the nuclear fuel cycle but in the wider fields of water chemistry, provision of pharmaceuticals, many metals, microelectronics and other materials.

This meeting on solvent extraction and ion exchange in the Nuclear Fuel Cycle will, undoubtedly, sharpen our thoughts on the present and future application of these two techniques in our own field of endeavour.

SOLVENT EXTRACTION AND ION EXCHANGE APPLICATIONS IN ORE PROCESSING AND URANIUM ENRICHMENT

Present and Future Use of Ion Exchange in Uranium Ore Processing

D. Naden, Davy McKee Research and Development
J.P. Pilot, Davy McKee (Nuclear) Ltd
C.W. Renner, Davy McKee (Stockton) Ltd

INTRODUCTION

Examination of uranium production over the past decade shows a steady increase to a record level of 44,000 tonnes per annum in 1981. This was followed by a slow decrease in demand and in 1983 production had fallen to 38,000 tonnes per annum

Most of the present plant capacity dates from the 1960's and the mid 1970's and the installed plant cost will therefore have been paid for by the high prices of the mid to late 1970's when uranium spot prices rose from $16 US/lb to $44/lb. The current spot price is less than $20 US/lb and there is therefore a restricted incentive to install new production capacity in situations where economics are the only consideration. Low prices and reduced demand, however, represent a challenge for technical development to reduce process costs and to increase profit margins. This challenge is being taken up and new techniques and processes are appearing over the whole range of production from ore sorting (eg high intensity magnetic separation) to new techniques and improvements in extraction processes (eg the combined mixer settler, CMS, and resin in pulp, RIP). Similarly new techniques of recovering uranium as a byproduct from existing process streams are being considered thus eliminating the costs of mining and beneficiation which can be seventy percent of total costs. Solid ion exchange has always found a place in uranium recovery and ion exchange developments have the greatest potential for simplifying the uranium recovery flowsheet and reducing recovery costs. It is this application of ion exchange which has the greatest interest to the uranium industry and which will therefore be examined in this paper.

THE IMPACT OF ION EXCHANGE ON URANIUM (IX) RECOVERY

Introduction

The sources of uranium and uranium recovery flowsheets have been adequately described elsewhere [1] whilst the selection of ion exchange equipment has also been the subject of various publications over the past decade [2] . The impact of ion exchange on the whole recovery flowsheet and process costs is reviewed in the following sections.

Recovery from Milled Ore

Uranium ore is usually in the oxide form and is finely disseminated in many ore bodies.

Milling to 90% less than 250×10^{-6}m (60 mesh) is often required to achieve rapid leaching by either sulphuric acid or sodium carbonate. Sulphuric acid is used where possible because of its efficiency in dissolving the ore to form soluble sulphate complexes with uranium and because of its low cost and environmental acceptability. When the acid consuming components of the ore are high then sodium carbonate leaching is used to form carbonate complexes of uranium which are also soluble. In both cases uranium is present in the leach pulp in dilute solution and both sulphate and carbonate complexes are readily taken up by both strong and weak base ion exchange (IX) resins from dilute solution.

The traditional post leach flowsheet has been (1) solid liquid separation and washing of leached ore to recover the dissolved uranium values, using either filtration or counter current decantation (CCD); (ii) solution clarification; (iii) uranium extraction by IX or solvent extraction (SX), or both, and uranium elution; (iv) uranium precipitation using sodium hydroxide, ammonia or H_2O_2. The first generation of SX and IX equipment required solution clarification in order to minimise organic losses in SX or packed bed blocking in IX. Packed bed ion exchange is still used in some of the older plants particularly in Spain and Canada. The drive towards reducing uranium production and plant operating costs, as well as the need for flowsheet simplication, led to the recognition of the need for an extraction process which could be used in the presence of suspended solids and hence eliminate the cost of solution clarification. Fluid bed IX was the first development to succeed in recovering uranium from solutions containing 1000 to 2000 kg/m^3 solids and equipment and processes using this technology have been widely adopted in Southern Africa [3] and, less widely, in North America. The most commonly used fluid bed continuous counter current ion exchange (CCIX) equipment is the Cloete-Streat column contactor [5], on which both the

NIMCIX and Davy McKee fluid bed CCIX designs are based, the Himsley column [6] and the Porter fluid bed tanks in horizontal arrangement [15]. SX engineering and equipment developments have subsequently led to the commercial acceptance of an SX process for treating solutions containing up to 500 ppm solids [4].

The post leach section of the flowsheet accounts for approximately 50% of the cost of the whole plant from ore crushing to product drying and handling whilst the solid-liquid separation and solids washing section accounts for over half this value (see Table 1). A more fundamental reduction in the capital cost may therefore be made by replacing all, or part of, the solid-liquid separation and solids washing step by a resin-in-pulp (RIP) recovery stage. The feed to RIP will depend on the ore type and the nature of the slimes stream. RIP technology has been used since the late 1960's in both acid and carbonate leach processes treating ore bodies with high clay content in Wyoming and New Mexico in the USA. The host rock treated generally consists of medium grain sandstone with calcareous cementing containing the uranium ore. The ores are crushed and milled to release the sand particles and after leaching the resulting particles are readily classified as sands and slimes in the ratio 70/30, the slimes being less than 50×10^{-6}m (325 mesh). The size and cost of the solids washing stage are controlled by the slimes fraction which is less easily washed on filters or in CCD settlers. The sands and slimes are therefore separated after leaching, the easily washed sands passing to low cost washing equipment such as cyclones or spiral classifiers. The slimes, together with the sands wash solution, form a pulp containing less than 5 to 10% solids which are passed to the RIP circuit in which stirred contactors are used with separation of resin and pulp between stages by external vibrating screens [1]. When sodium carbonate leaching is practiced the slimes are particularly difficult to filter and wash and RIP in stirred contactors is therefore especially acceptable. The pulps produced from these ores do not generally lead to high resin attrition losses and resin make up is approximately 30% of the inventory/annum. This level of resin loss contributes less than 2c/lb U_3O_8 and is therefore considered to be acceptable.

The wider application of this process route to sandstone ores which do not have a high clay content, and therefore produce pulps with both low viscosity and low density, using fluid bed CCIX has been suggested [2]. Plant capital cost savings of up to 25% were indicated [7]. The feasibility of using fluid beds CCIX was established [8] [9] and one plant is reported to be operating using 0.5% solids pulp at a uranium mine in France. The process has not however been widely accepted.

When the uranium ore is dispersed in a quartz conglomerate host rock, as in most Canadian and South African ore bodies, milling to 90% less than 60 mesh before leaching is usually practiced. The milled ore readily settles and filters, and solid liquid separation and washing are not therefore difficult in this case. RIP, although often proposed, has never been commercially used. Unlike the sandstone ores there is no sharp cut off between sands and slimes and separation will therefore be selected for flowsheet convenience. In general the solids content of an RIP feed stream will be higher than in the case of sandstone ore treatment. Typically an ore milled to 60 to 70% below 200 mesh will provide an RIP feed containing 25% suspended solids when the split is carried out at 200 mesh. In some cases however the whole leach product containing 50 to 60% solids can be passed to RIP. Milled quartz ores are more abrasive than sandstone ore slimes streams and resin losses can be expected to be somewhat higher than in present RIP plants.

In South Africa uranium is often produced as co-product with gold and the ore mining and milling has already been carried out. Carbon-in-Pulp (CIP) is now becoming widely practiced in South Africa and the use of RIP for uranium recovery is therefore the next stage in the simplification of the whole process flowsheet. This is further discussed in the case study in Section 3.3.2.

In-Situ Leach

In-situ leach (sometimes called borehole solution mining) and stope leaching have the advantage of lower capital and operating costs than conventional mining as in both cases ore mining and transport to the surface, ore comminution, leaching in vessels and tailings disposal are eliminated. Both acid and carbonate leaching have been used and solution containing little suspended solids is produced with uranium concentration ranging from 0.02 to 0.2 kg U_3O_8/m^3. Solution flows are typically 5 to 10 m^3/minute [10]. Solid ion exchange recovery is exclusively used as the solutions are dilute (thus precluding direct precipitation) and, unlike SX, resins will not contaminate the raffinate which is recycled to the borehole. When carbonate leach is practiced recovery by resins is the only feasible process owing to the adverse chemistry of the SX process. In-situ leach has been used in the sandstone deposits in Texas (eg at the Arco Clay West Plant) while stope leaching was briefly used at the Kerr-Addison Agnew Lake Mine [11].

Byproduct Uranium

Uranium is a byproduct when most of the ore mining and preparation, including leaching, have already been carried out for another primary product. These, although profitable to the producer are often minor sources. Recovery from copper ore leach liquors containing up to 15 kg U_3O_8/m^3 has been practiced using IX processes at Twin Buttes and Kennecott in the USA. Mine drainage waters, underground slimes streams and heap leaching may also be included in this category. A survey by the USBM of 14 acid and carbonate mine waters indicate that these contain 1 to 12 kg/m^3 [12].

As already stated uranium may be considered as a co-product when recovered from South African gold ore pulps as in this case the uranium must support the separate leach and solid liquid separation.

The principal potential source of byproduct uranium has always been phosphoric acid produced by sulphuric acid leaching of phosphate rock. The U_3O_8 content of these ores from a number of locations worldwide ranges from 0.005 to 0.02% (see Table 3) which can lead to 0.10 to 0.2 kg m^3 U_3O_8 in solution. This compares favourably with that of in-situ leach streams in the USA (eg 0.16 to 0.2 kg/m^3). It is estimated that 2.5 x 10^6 kg U_3O_8/annum is available from this source in the USA alone.

Both SX and liquid membrane (LM) processes have been developed for uranium recovery from phosphoric acid but at present uranium prices these have limited application owing to high process costs. Recently however resin processes have been investigated by Streat and Gonzalez-Luque at Imperial College, London using both novel and commercially available resins [13] and several resins have been identified as having industrial potential. The initial investigation has been followed up by an SERC support programme at Imperial College, sponsored by Duolite International and Davy McKee, in which a resin was further investigated. Simultaneously a pilot plant activity on a phosphoric acid stream in Israel by Ketzinel, Volkman and Hassid of the Israeli Nuclear Energy Commission [14] has demonstrated a similar process and it is claimed that uranium can be produced at approximately $44/Kg, which is half the cost of SX process costs and is near to the present spot price. A resin ion exchange process therefore has considerable potential for recovering uranium from this source.

THE INFLUENCE OF IX DEVELOPMENTS ON FUTURE RECOVERY PROCESSES

Introduction

The potential of solid ion exchange to significantly reduce uranium recovery cost is unique and is due mainly to the insolubility of resins. This enables resins to be used for recovery from solutions containing a high level of suspended solids and avoids contamination of the process solution. The development of improved resin chemistry for specific types of processes can also be expected to influence the use of resins in uranium recovery.

Fluid Bed Ion Exchange

Fluid bed technology, in both columns and separate tanks in horizontal arrangement, treating unclarified solutions, is now mature and is widely used in the uranium industry. Twin columns each having twelve stages and a diameter of 4.85 metres are in use at Chemwes in South Africa treating 270 m^3/minute of solution [15]. Tanks with 6m side length treating 850 m^3/h are in operation at Rossing Uranium Ltd [16]. There is no theoretical limit in either case to scale up. These operations have been described elsewhere while the influence on the uranium recovery flowsheet is also well documented [1][7]. The Cloete-Streat, or NIMCIX, system is the most widely used column system and the Porter system the most widely used open tank system. Operating experience has been good and minor modifications to ancilliary equipment have ensured smooth operation following minor early problems. The only major weakness of fluid bed ion exchange columns which remains is the accumulation of solid particles on the column plates. In some situations this has led to compacting of resin and ore solids which have proved difficult to remove on back-flow. Several means of solving this problem without the use of external equipment present themselves and a simple solution would enable the column to be much more reliably used for a wider range of feeds with a higher solids content (in some cases up to 10% solids[8][9]).

Resin-in-Pulp (RIP)

Improved RIP Process

The development of air swept screens for carbon-in-pulp (CIP) recovery of gold by MINTEK in South Africa [17] and the successful operation of this technology by the South African gold industry has led to the much wider application of in-pulp technology to feeds with 50% solids content. Air swept screens for CIP are now well established and have also been piloted in South Africa for RIP recovery of gold. Davy McKee have developed an improved deep submerged air-swept screen design which will operate in the presence of much higher concentrations of both carbon and resin than in CIP or RIP processes currently being operated [18]. The result is a significant reduction in plant size and greatly improved ease of operation. The development of submerged air swept screens has also enabled the whole RIP extraction system to be completely redesigned. Separate agitated vessels with external

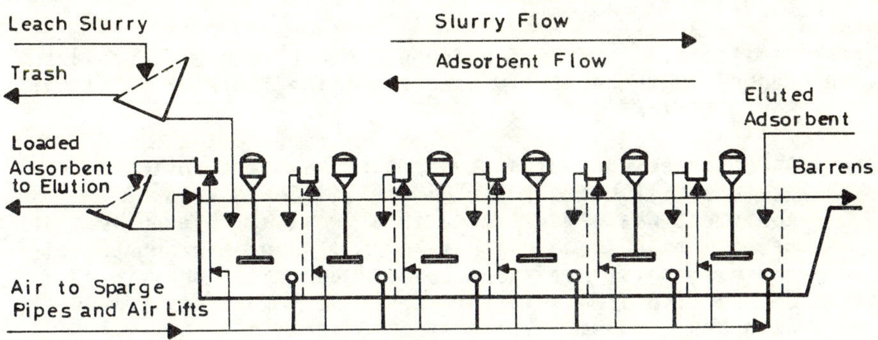

FIGURE 1
Davy Mckee Improved Resin-in-Pulp Contactor Design (Diagramatic)

vibrating screens are no longer required as separation of resin and pulp can be carried out by placing an air swept screen in the pulp flow between connecting stages. The resulting RIP contactor design is illustrated diagramatically in Figure 1. The development of the new Davy Mckee RIP technology was carried out in the following systematic schedule of activities:

i) A UK based hydraulic programme carried out in a two stage test rig with a capacity of 7 m^3/hour in which the screen design was developed and screen operating data obtained.

ii) A bench scale programme in which a range of resin types were surveyed.

iii) Modification of the continuous counter-current modelling technique successfully used to design fluid IX plants in South Africa.

iv) An engineering design study which resulted in the type of equipment illustrated in Figure 1. The developed equipment was demonstrated in a two stage hydraulic pilot plant with a stage capacity of 8 m^3 and a throughput of approximately 75 m^3/hour pulp. Operating conditions covered those anticipated when treating pulps from both acid and carbonate leach of both sandstone and conglomerate ore bodies, ie 10% to 50% w/w solids. Operating data was obtained with resin concentrations of up to 25% v/v.

A comparison of the size and cost of conventional and Davy McKee in-pulp contactors for an ore throughput of 100,000 tonne/month as a 50% slurry has been carried out and the reduction in size is illustrated in Figure 2. The resulting size reduction leads to a capital cost reduction of approximately 45%. In addition the ease and reliability of operation is significantly improved. Smaller equipment size also means that building costs are reduced when the climate requires that the plant must be housed in buildings.

RIP Recovery of Uranium in a Combined Gold/Uranium Mill

An investigation was carried out by Davy McKee of uranium recovery from an underground slimes stream containing 40% w/w solids following a cyanide leach and carbon-in-pulp recovery of gold. A programme of bench scale leach and RIP modelling tests was followed by a pilot plant programme in which RIP hydraulic data was obtained. The RIP design date was confirmed in the on-site programme using 50% w/w solids pulp.

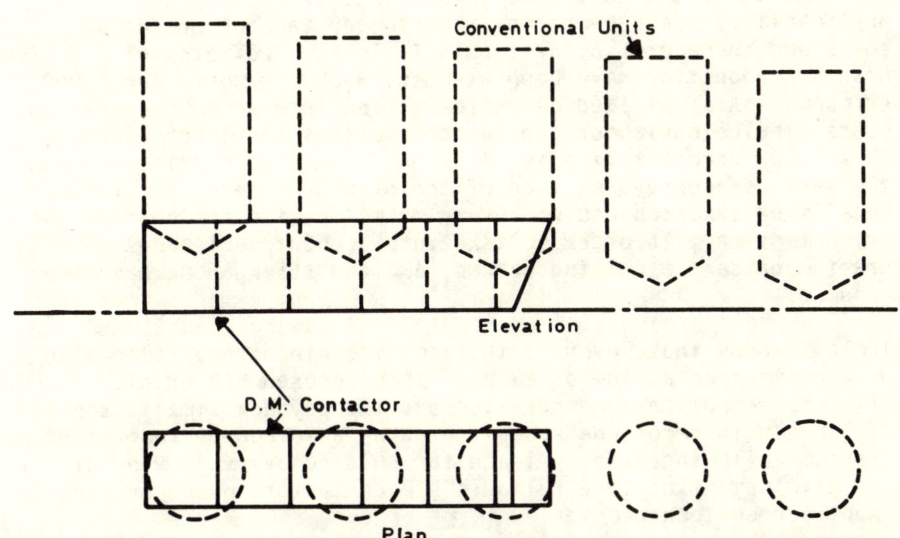

FIGURE 2

Comparison of Conventional and Davy Mckee (DM) Improved Resin-in-Pulp Contactors for 100,000 tpm Ore as a 50% Slurry (assuming 5 stages only)

The capital and operating costs have been estimated for uranium recovery from 10,000 tonnes/month of slimes in a process involving acid leaching, RIP recovery of uranium from the leach pulp, resin elution and regeneration, SX recovery from the IX eluate, SX strip and ammonium diuranate precipitation. A problem associated with acid uranium leaching following gold leach by cyanide is the need to either destroy cyanide or to treat the HCN produced and this entailed additional capital and process costs.

The results of the cost study, modified for wider applicability, have been converted to capital and operating costs and these are summarised in Table 2. Two rates of uranium production have been assumed, 8,000 kg/month and 5,000 kg/month. A simplified cash flow taking into account operating costs (including labour and maintenance) of the leach and RIP, SX and ADU precipitation sections has also been carried out and the revenue and payback time of the plant has been calculated. Details of taxation etc will vary from location to location and are not given. In order to take into account variations in uranium price two selling prices, \$33 and \$55/kg U_3O_8 have been assumed.

Table 3 shows that, even at this small scale of operation with a uranium price as low as \$33/kg, the proposed RIP uranium plant is economically attractive and the payback time is short. RIP should therefore be reconsidered as a method of recovering uranium following cyanide leach for gold recovery in view of the development of more reliable RIP contactors based on air swept screen technology.

RIP Recovery of Uranium from Sandstone Ore Slimes Stream

The replacement of slimes washing in the recovery of uranium from sandstone ores by a resin in pulp (RIP) process can result in a capital cost saving of up to 50% in the post leach section of the flowsheet. The resulting saving in the capital cost of the whole mill can be 25%. Equipment currently in operation is large and has fairly high operating and maintenance costs, in addition the use of external vibrating screens for separating resin and pulp presents a limit in scale-up and applicability. RIP is therefore considered at present only for ores with a high clay content which are difficult to treat by other means. The use of submerged in-situ air swept screens will enable the wider use if RIP in future and the potential savings in capital cost may be realised for a much wider range of ores than those for which RIP is currently considered. The development by Davy McKee of an improved air swept screen which will operate in the presence of high resin concentrations and the improved engineering design of the compact, low profile RIP extraction unit has already been described. The significantly improved ease and reliability of operation of this type of RIP contactor will enable RIP to be considered in all future flowsheets in which sandstone ores are treated and will therefore enable resin ion exchange to make its maximum contribution to uranium recovery costs.

Uranium Recovery from 30% P_20_5 Phosphoric Acid

Uranium not recovered from phosphoric acid is redispersed with the fertiliser product and, although this does not represent any hazard to the environment, it does represent a lost resource. There is therefore considerable incentive to recover uranium in countries where phosphate ores represents a major national resource and in countries which have no uranium resources but which import phosphate rock or phosphoric acid.

A comparison of the cost of a resin IX process with the cost of a solvent extraction proccess is given in Table 4. Much of the saving associated with the resin IX process is due to the elimination of acid pretreatment, aimed at removing SX crud forming materials, and acid post treatment, aimed at removing SX organic reagents which would damage down stream evaporators rubber linings in the phosphoric acid plant. It is assumed that neither of these steps are required in a resin ion exchange process. The resin ion exchange process is thus expected to make a significant impact on uranium recovery from this source. Following the bench scale investigation at Imperial College in which data was obtained on recovery from a number of acids an outline flowsheet has been designed [13] and a preliminary budget estimate has been made of the capital and operating costs based on the flowsheet concept. The results for an acid containing 150 ppm uranium are summarised in Table 5. The resin used, extracts a wide range of other metal ions as well as uranium. The relative extraction of metals is dependent on their concentration and their oxidation state, a sufficiently good uranium selectivity and level of loading was however achieved in the development programme for the process to be considered as commercially attractive. The resin loading assumed for the cost study is based on that achieved in both the Imperial College programme and in a separate study carried out by Davy Mckee Research and Development. Uranium is eluted from the resin together with some other ions and a dissolution and re-precipitation of uranium will therefore be required. Conventional resin handling techniques are assumed throughout and a resin make-up cost has been included to take account of losses and deactivation. No information is however yet available on the effect of the build-up of minor impurities on the resin which may result in temporary or permanent loss of resin activity.

Although the recovery and recovery costs will vary from acid to acid, the preliminary estimate of approximately \$53/kg of U_3O_8 for a 110,000 tpa P_2O_5 plant is sufficiently attractive to continue the development of this process. The cost for larger plants will be even more attractive and assuming the estimate cost savings are maintained the process could result in a major new source of uranium from an old resource.

The initial work must now however be followed up by an on-site demonstration pilot activity using fresh acid to obtain data for a full scale plant design.

CONCLUSIONS

Resin ion exchange in recovering uranium from a wide range of solutions is well established although the large potential impact of ion exchange on uranium recovery flowsheets has yet to be realised. Developments in both resins and equipment have now reached the stage however when the realisation of this potential can be anticipated in the near future. The two applications identified in this paper, the resin process for uranium recovery from phosphoric acid and the new compact, high throughput resin-in-pulp process can be expected to both significantly improve recovery costs and simplify the recovery process during the next few years.

ACKNOWLEDGEMENTS

The authors would like to thank both Davy McKee Management for permission to publish this paper. The development of the Davy McKee in-pulp technology was supported by funding from the Department of Trade and Industry and was greatly assisted by Rand Mines Ltd of South Africa who provided a site for the demonstration of the technology. In addition appreciation is expressed to the Science Research Council, Imperial College, phosphoric acid producers and the management of Davy McKee Research and Development for providing funding, facilities and materials for the investigation of the resin ion exchange process for the recovery of uranium from phosphoric acid.

REFERENCES

[1] Merritt R C, "The Extractive Metallurgy of Uranium" Colorado School of Mines Research Institute

[2] Naden D, Bandy M R, "Choice and Design of Solid Ion Exchange Plants for the Recovery of Uranium" J Chem Tech Biotechnol, 1979, 29, 232-248

[3] Boydell D W, "Bufflex and Related Circuits for the Recovery of Uranium", National Institute for Metallurgy

Symposium, Ion Exchange and Solvent Extraction in Mineral Processing, NIM Randburg, February 1980

[4] Rowden G A, "Davy McKee's Combined Mixer Settler", Chem E Symposium, Liquid-Liquid Extraction Science, Extraction '84', Dounrey, Scotland, November 1984

[5] Cloete F L D, Streat M, British Patent 1,070, 215 (1982)

[6] Himsely A, Canadian Patent 980, 467 (1973)

[7] Naden D, Willey G, Newrick G M, "The Use of Fluid Bed Ion Exchange to Reduce Uranium Recovery Costs", 105 AIME Annual Meeting, Las Vegas, February 1970

[8] Streat M, Qassim R Y, "Recovery of Uranium from Unclarified Liquors by Ion Exchange" 120th International Symposium on Hydrometallurgy, Chicago, USA February 1973

[9] Ritcey G M, Slater M J, Lucas B H, "A Comparison of the Processing and Economics of Uranium Recovery from Leach Slurries by Continuous Ion Exchange and Solvent Extraction" 120th International Symposium on Hydrometallurgy, Chicago, USA, February 1976

[10] "In Situ Leaching Opens New Uranium Reserves in Texas" E/MJ July 1975, p3

[11] "Solution Uranium Mine to Start Production", Mining Magazine, September 1977, p191

[12] George D R, Ross J R, "Recovery of Uranium from Uranium Mine Waters and Copper Ore Leaching Solutions", Processing of Low Grade Uranium Ores, Vienna 27 June to 1 July 1966

[13] Gonzalez-Luque S, Streat M, "The Recovery of By-product Uranium from Wet Process Phosphoric Acid Solution Using Selective Ion Exchange Resins", Ion Exchange Technology, Edited by D Naden and M Streat, Published by Ellis Horwood Ltd Chichester, England 1984

[14] Ketzinel Z, Volkman Y, Hassid M, "Research on Uranium Recovery from Phosphoric Acid in Isreal - Progress and Prospects" Advances in Uranium Ore Processing and Recovery from Non-Conventional Resources, IAEA, Vienna, September 1983

[15] McIntosh A M, Viljoen E B, Craig W M, "The Design, Commissioning and Performance of the NIMCIX Section of the Chemwes Uranium Plant" National Institute for Metallurgy

Symposium, Ion Exchange and Solvent Extraction in Mineral Processing, NIM Randburg, February 1980

[16] Vernon P N, Sylvester C W, "The Rossing Continuous Ion Exchange Plant" National Institute for Metallurgy Symposium, Ion Exchange and Solvent Extraction in Mineral Processing, NIM Randburg, February 1980

[17] Laxen PA, Becker G M, Rubin T, "The Carbon-in-Pulp Gold Recovery Process - A major Breakthrough", South African Mining Engineering Journal, July 1979, 40, 4152, 43-59

[18] Naden D, Willey G, Bicker E, Lunt D, "Development of an In-Pulp Contactor for Metals Recovery" Ion Exchange Techology, Edited by D Naden and M Streat, Ellis Horwood Ltd, Chichester, England 1984

TABLE 1 Uranium Recovery from Milled Ores, Plant Capital Cost Breakdown for SX or IX plant with Solution Clarification [1]

	Approx %
Ore handling, crushing and sampling	20
Ore storage, grinding and density control	15.5
Leaching	10
Solid Liquid Separation and solids washing	31
IX or SX	15
Precipitation, filtration, product dry and handling	7
Tailings Disposal	1.5
TOTAL	100.0

TABLE 2 Costs of Uranium Recovery from Slimes (January 1985)

Ore throughput - 10,000 tonnes/month

Uranium Recovery

a) 8,500 kg/month

b) 5,000 kg/month

Capital costs	SA Rand x 10^6
Leach Plant (including HCN recovery)	1.82
RIP Plant (including resin)	2.27
SX and ADU precipitation (including SX reagent)	1.70
Total Capital Cost	5.79

Operating costs	SA Rand/tonne ore
Leach plant	7.37
RIP and SX	2.09
Downstream costs (including lime for neutralisation)	1.18
Total Operating Costs	10.64

U_3O_8 selling price		$US 33/kg$U_3O_8$	$US 55/kg U_3O_8
Cash Flow (Rand)			
Monthly revenue	(a)	596,800	994,400
	(b)	373,000	621,500
Monthly operating cost		106,400	106,400
Net monthly revenue	(a)	490,400	888,000
	(b)	266,600	515,100
Capital payback time - months	(a)	11.8	6.5
	(b)	21.7	11.2

2.26 Rand/$ exchange rate (Financial Times 15.1.85)

TABLE 3 Typical Uranium Content of Phosphate Rocks for Various Locations

Rock Source	U content of rock kg/tonne	U content of 30% P_2O_5 acid kg/m^{-3}
Morocco	0.11	0.135
USA (Florida)	0.11	0.135
Togo	0.10	0.125
Tunisia	0.06	0.075
Senegal	0.16	0.200
Jordan	0.12	0.150
Israel	0.12	0.150
Algeria	0.10	0.125

TABLE 4 Comparison of Equipment Costs for a Typical SX Process and a Proposed IX Process for Recovery of Uranium from Phosphoric Acid

	Conventional 2 Step SX Process %	Proposed Resin IX Process % of SX Total Process Cost
Feed acid clean-up (mainly clarification)	27.6	14.2
eH adjustment, oxidation and reduction	1.6	2.8
Extraction or elution	30.8	19.9
Post extraction acid clen-up	21.1	-
Uranium precipitation etc	18.9	18.9
Total	100.0	55.8

TABLE 5 Uranium Recovery from Phosphoric Acid by Resin Ion Exchange Process (based on Data from the Experimental Programme)

Plant Details	
Phosphoric acid plant capacity	110,000 tpa P_2O_5
Acid strength	28%
Uranium content of phosphoric acid	0.150 kg/m^3 as U
Uranium recovery	95%
Uranium Production/Annum	49,500kg of U_3O_8
Cost of U_3O_8 Production, Mid 1984 £ Sterling	
Capital cost	2.50 x 10^6
Start-up (including first chemical charge)	0.25 x 10^6
Total Capital	2.75 x 10^6
Operating Cost/Annum £ x 10^6	
Labour	0.128
Maintenance 5% of capital cost	0.125
Chemicals and utilities	0.936
Resin make up	0.055
Total Direct Operating Cost	1.244
Capital Cost Contribution £ x 10^6	0.820
Capital Depreciation £ x 10^6	0.230
Total Production Cost £ x 10^6	2.294
Cost Contribution to U_3O_8 Production	46.3 £/kg U_3O_8 21.0 £/lb U_3O_8
Cost in $US assuming $1.15 - £1.0	
	$53.2/kg U_3O_8
	$24.2/lb U_3O_8

Solvent Extraction in Uranium Ore Processing

Gordon M. Ritcey, Head, Solution Purification Section CANMET, Energy, Mines and Resources

INTRODUCTION

The recovery of uranium from ores by using solvent extraction began in 1955 with the use of DEHPA (DAPEX Process) [1], and since 1957, the tertiary amines (Amex Process) [2] have been the most popular extractants.

The most common lixiviant for uranium ores has been sulphuric acid, although at least one plant has used nitric acid with a TBP extraction [3], and a few plants have used alkaline leaching with direct precipitation of the yellowcake [4]. Several plants use a combination of ion exchange with solvent extraction (Eluex or Bufflex process) in treating sulphuric acid leach solution containing uranium [5,6].

For the most part, mixer-settlers have been the contactor used in the solvent extraction processing of uranium-bearing liquors, and many modifications and designs have been presented since the early gravity mixer-settler.

This paper will describe the role of solvent extraction in the processing of uranium ores up to the present and possible future applications of solvent extraction in developing technology.

The reader should refer to certain texts such as Clegg and Foley [7] which describes the industry up to 1958, Merritt [8] up to 1971; Ritcey and Ashbrook [9] which details the subject up to 1977; numerous proceedings of international uranium agencies; and the proceedings from several

International Solvent Extraction Conferences held in Gatlinburg (1962), Harwell (1965), Gothenburg (1966), The Hague (1971), Lyon (1974), Toronto (1977), Liege (1980) and Denver (1983).

EXTRACTANTS AND MODIFIERS

The tertiary amines continue to be the most widely used for treating process solutions resulting from sulphuric acid leaching of uranium-bearing ores. The major impurity that is extracted preferentially to uranium is molybdenum, and it can be removed by carbonate stripping after first selectively stripping the uranium with a chloride solution. Oxidized vanadium if present, most of which will be co-extracted and stripped with molybdenum [10]. Zirconium extraction can be suppressed by the addition of phosphate. Iron is water scrubbed and any arsenic present may be removed in a scrub stage [11]. Feed solutions having high chloride levels present, as in the case where the solution results from chloride elution of an ion exchange resin, may be treated using an alkylphosphate, amine (after dilution of the feed) or by TOPO [12]. The use of TOPO as a synergist with the tertiary amine provides for increased kinetics. The stripping is difficult, but can be accomplished with an H_2SO_4 + H_2O_2 mixture [13]. The TOPO replaces the long chain alcohol as a modifier in such a system. As for modifiers, there have been no satisfactory replacements for the long chain alcohols. However, the use of TOPO for treatment of phosphoric acid has been successful, and one can anticipate its use in the future for recovery of uranium from sulphuric acid solutions.

DILUENTS

For many years the diluent was considered to be only the carrier for the active extractant, with no other function in the extraction system. This myth has been dispelled over the past few years and indeed the diluent does play an important part in the extraction process, with aromaticity, solubility parameter, naphthene content, etc. exhibiting effects [14]. Diluents other than the kerosene type have been investigated, particularly perchloroethylene which is heavier than water [10,15,16]. With the latter type, the phase separation is usually faster, and metals discrimination isoften improved as compared to the aliphatic and aromatic diluents. Flash point is an important criterion in the choice of a diluent, and most plants select one with a flash point of greater than 120°F. Because of the diluent composition affecting the solvency and chemical properties of the extraction, some diluents are now supplied by the manufacturer in a blended form to contain 15-20% aromatic content in an aliphatic diluent. Such a diluent is the Escaid 100 by Exxon. The diluent should be selected only after sufficient tests have indicated it to be compatible with the process.

SCRUBBING

In the early uranium operations in the 1950's and 1960's the solvent extraction plants were designed to include a scrub circuit prior to stripping for uranium recovery. The purpose of the scrub circuit was to remove any entrained or sorbed impurities that might be carried with the loaded solvent and which could adversely affect the subsequent stripping and phase separation characteristics. The impurities were generally Ca, Mg, Fe, Al, Si [17]. Water was found to be a satisfactory scrub solution. However in the second generation of uranium plants there is a high percentage of circuits not incorporating the scrub stage. The results of course have been poor phase disengagement, high solvent losses due to entrainment in the aqueous phase, crud formation and losses of solvent, and in general increased operating cost. By examining the causes of emulsion formation in the stripping circuit a suitable scrub stage can be incorporated to improve the stripping circuit efficiency.

STRIPPING

Much of the emphasis in solvent extraction development research on new processes is directed towards the extraction circuit, with minimum consideration often given to the stripping circuit. In the uranium circuit the stripping agents have included the following: Na_2CO_3, NaCl, NH_4Cl, Na_2SO_4, $(NH_4)_2SO_4$, $(NH_4)_2CO_3$.

Because of the poor pH control often encountered when using ammonia addition, uranium precipitation can occur within such circuits. The result is the formation of excessive cruds and high solvent losses. This reflects adversely on the economics of the plant. The stripping technique should be chosen to be compatible with the following: the extractant and possible degradation; ultimate product desired; impurities specifications in the final product; co-extracted impurities that could precipitate or hydrolyse during stripping; the ease of pH control during stripping; the design of the contactor selected and therefore the phase separation characteristics; equilibrium efficiency; and the possibility of recycle of the stripping solution following precipitation of the final product.

Apart from a few plants that employ the chloride stripping in the presence of co-extracted molybdenum [8], the majority of the uranium circuits use controlled pH stripping with $(NH_4)_2SO_4$ and ammonia addition. For example, ammonia-ammonium sulphate stripping is used in Canada, Australia, United States and South Africa. Sodium chloride stripping is effectively used in mills in the United States, France, Niger and Gabon [18]. One plant in the United States and one in Niger use sodium carbonate stripping.

CONTACTORS

The extraction circuit must be designed to ensure that the proper selection of the equipment has been made, and that optimum operation of the equipment is maintained in order to provide maximum mass transfer efficiency with minimum solvent losses. There is no universal contactor and the chemical and physical characteristics of the systems relative to dispersion and coalescence, kinetics, and mass transfer efficiency must be considered in the choice of the contactor [10].

The mixer-settler has been the most widely used contactor in uranium processing, originating with the gravity type design, followed by the pump-mix design and its many modifications in pump design and settler configuration [9]. The mixer settlers are claimed to be ideal as they are inexpensive, easy to fabricate, usually easy to cleanse of entrained solids or cruds, easily scaled-up and simple to operate and control. However they do have a large solvent hold-up, occupy a large area, which can be expensive if requiring a heated building, and the combined pumping-mixing provides an extremely high shear to the process solution and solvent mixture and so droplets are produced which are often much smaller than necessary to effect the required mass transfer. Consequently stable emulsions, and often cruds, are produced.

The performance of gravity settlers has been improved by the use of vertical baffles and picket fences [19,20) and the general aspects of mixer settler design have been well documented in recent years [21-24]. In addition to baffles to improve coalescence, work on packing, or coalescing material has increased the capacity of settlers [24].

Many plants using high shear types of mixer settlers must clarify the feed to less than 30 ppm solids to minimize crud formation. In the past few years there has therefore been more concern as to the generation of crud, its cost in solvent loss, as well as the concern for the environment receiving the solvent discharged to tailings. The crud often contains up to 80% or more of organic phase. Solvent losses, and those attributed to crud, can be caused by many variables in the process, both chemical and physical [9,25-30], and various crud treatment methods have been proposed [26].

The equipment manufacturers now have a better appreciation of the problem of high shear and stable emulsions and cruds, and the designs of the past couple of years have demonstrated equipment having minimum shear, but sufficient to provide for mass transfer. The Krebs mixer-settler [31,32] is gaining fast acceptance since 1975 with uranium plants incorporating the unit installed in Gabon, France and South Africa, and Canada. The design permits for 3-4 times the flow capacity of the usual pump-mix design and could probably tolerate more solids in the feed solution.

Davy McKee have developed a contactor, the Combined Mixer Settler (CMS) [33-34] to minimize turbulence. It can accommodate up to 300 ppm of solids without generation and accumulation of crud. Davy McKee have been awarded contracts to install plants at flow rates of 250 m^3/h and 700 m^3/h [33]. A capital cost saving of about 30% over conventional mixer settlers is claimed, together with lower operating costs.

SOLVENT RECOVERY

Solvent losses and recovery are documented in the literature [9]. However, an interesting approach to solvent recovery from the raffinate is to wash with the diluent [35,36]. Mixer settlers or centrifugal contactors can be used. The system reported was TBP-HNO_3 but amine-H_2SO_4 should be equally viable. Activated carbon and various types of coalescers have been tried. One design, demonstrated on kerosene dispersed in water consisted of a fibrous medium treated with isobutyl methacrylate resin to form a rigid and stable structure [37]. Regeneration of the coalescer is by steam. Solvent can also be recovered by the use of electrostatic coalescence [38]. C-E Natco's demonstration on the circuit at United Nuclear's Church Rock, New Mexico uranium mill, showed that 5 times as much feed could be processed per unit cross section of vessel area as compared to that of a conventional mixer settler [38].

SOLVENT IN-PULP

If substantial reduction in processing costs are to be achieved in the mill, then the liquid-solids separation must be eliminated. This costly operation accounts for approximately 50 percent of the capital and operating costs of the mill. Such a possibility exists in the use of resin-in-pulp continuous ion exchange or solvent-in-pulp extraction from the leach slurry [9]. The Beijing Research Institute of Uranium Ore Processing in China is presently engaged in considerable research on solvent in-pulp, using sieve plate pulse columns [39] and other equipment design. A plant is in operation using columns. CANMET has also returned to research in this important area.

BY-PRODUCT RECOVERY

Most hydrometallurgical process plants ignore the fact that there are potentially valuable by-products that might be recovered in the process. These values unfortunately are often lost and impounded with the tailings, usually because the market demand is not sufficient at that time. For example, in Canada the uranium ores in the Elliot Lake area also contain considerable amounts of thorium and rare earths [40], and in the high grade Saskatchewan ores there are substantial amounts of arsenic and nickel [11]. In mills in the USA, Brazil and elsewhere, molybdenum has been recovered as a by-product in the uranium circuit by stripping with carbonate after first selectively stripping the uranium with NaCl [8].

A process has been developed by the US Bureau of Mines to recover vanadium as a by-product [41]. Following uranium recovery, vandadium is extracted from the uranium raffinate with either the alkylphosphoric acid D2EHPA or the alkylphosphinic acid Cyanex 272 after first adjusting the raffinate pH to 2 and reducing the emf to less than 115 mV with iron powder.

ENVIRONMENT

Although the development and optimization of the metallurgical processes, including solvent extraction, has progressed through phases of concern for economics and energy, a more recent area of concern is for industrial plant hygiene and the environment. Therefore, as noted under equipment, the proper selection of the contactor design and proper operation is necessary to minimize solvent losses in the plant and in the effluents discharged. Poorly vented work areas and the use of volatile solvents cause hygiene concerns in the plant but these can easily be eliminated. Waste raffinates, often containing large amounts of entrained solvents, may be discharged to the environment if the equipment and its operation are not optimal. Future environmental constraints may well force the operator to provide for in-plant treatment of such raffinates to minimize the environmental impact. Coalescers, centrifuges or perhaps activated carbon treatment may be necessary to control the amount of organic released to the environment [42].

Recent studies have been carried out at CANMET to determine what could be expected in the tailings over a period of time if organic constituents from the uranium solvent extraction operation are discharged [43]. The tests were conducted under controlled, accelerated weathering conditions in lysimeters. The results of the tests indicated that the amine was washed out of the tailings after about two years and the kerosene after 11 years. The isodecanol was difficult to analyse but appeared to follow the kerosene.

NEW ALTERNATIVE TECHNOLOGY

If we are to consider the development of alternative technology, then the technology selected must possess many advantages. A list of such desired objectives are presented in Table 1. Obviously it would not be necessary to fulfill all the stated requirements, but if the environmental problems and uranium recovery criteria are satisfied then any other benefits would be additional.

TABLE I. Emphasis and rationale for considering alternative flowsheets

- Decrease the environmental impact
- Isolate radium in mill
- Increase the uranium recovery (95-98%)
- Increase the possibility of by-product recovery (As,Ni,Th, RE's)
- Decrease the capital and operating costs (eg. liquid/solids separation)
- Decrease energy requirements
- Increase solution recycle
- Decrease acid drainage in tailings
- Decrease overall tailings maintenance

The following discussion compares some of the flowsheets resulting from research that has been ongoing at CANMET during the past 5 years on improvements in the uranium processing flowsheet.

Before examining recent research and considering alternatives to the present process, we must first look at the advantages and disadvantages of the present technology. Table 2 notes some of those major factors, the main disadvantage being the aspect of tailings management.

Table 2. Conventional flowsheet (H_2SO_4)

Advantages

Proven over time
Produces acceptable uranium product for further refining
Some by-products are possible

Disadvantages

High cost L/S separation (capital & operating)
High cost tailings management
400 pCi/g Ra^{226} in tails (based on approximately 0.1% U_3O_8 in ore)
No solution recycle (essentially no provision)
High lime requirements for neutralization

CHLORINATION ROAST-LEACH PROCESS

Because H_2SO_4 does not solubilize the radionuclides, other circuits were investigated. Using HNO_3, most of the radionuclides were solubilized when treating ore containing 0.1% U_3O_8. Nevertheless the tailings still contained up to 100 pCi/g Ra^{226} [44,45]. Further studies included the possibility of using the chloride ion to provide radionuclide solubility. Two routes were concurrently developed; one a chlorination roast followed by leaching, the other a leach in HCl. This was a result of an extensive literature search where possible routes were cited [46-51].

The chlorination roast process consists of a two-stage chlorination, at 400°C to volatilize the uranium chloride, followed by 600°C to convert the radium to a chloride. The residue is then leached in dilute HCl to solubilize the radium chloride, as well as any uranium, thorium and rare earths that may be present. This solution is then treated to remove and isolate the radium for burial and a solution bleed is sent to the uranium and thorium recovery circuit. The major portion of uranium from chlorination would be condensed and the condensate purified to recover uranium, thorium and remove iron. Chlorine would be recovered in the process to be recycled to the roasting stage. The process has been demonstrated in a small continous shaft furnace of 0.5-1.0 kg/h capacity, and residues containing 15-20 pCi/g Ra^{226} were produced. In Table 3 the advantages and disadvantages of such a process are noted.

Table 3. Chlorination

Advantages

$<$ 25 pCi/g Ra^{226} in tails on low grade ores
Increased by-product recovery possible
Decreased sulphides in tails
Much of solution recycled
High purity uranium product possible
Ra^{226} isolation in mill
Increased uranium recovery

Disadvantages

Cl_2 recovery necessary for economics (good engineering required)
Higher maintenance likely
Unproven on complex ores (200 pCi/g Ra^{226} the lowest achieved)
High costs on low grade ores
Integrated circuit has not been run

HYDROCHLORIC ACID LEACH PROCESS

In the hydrochloric acid leach process, the ore is leached in HCl with the addition of sodium chlorate as an oxidant [53]. Either a one or two-stage leach at 70-80°C is employed, depending upon the ore, giving a final tails containing about 10 pCi/g Ra^{226}. The uranium would be recovered by solvent extraction using trioctyl-phosphine oxide, stripping with H_2SO_4 + H_2O_2 mixture followed by precipitation of uranyl peroxide with H_2O_2 [54]. The thorium would then be recovered by solvent extraction and the radium removed and isolated on an ion exchange resin. The solution is returned to leaching, with the addition of HCl and $NaClO_3$. The advantages and disadvantages of such a process are given in Table 4.

Table 4. HCL leaching

Advantages

$<$ 25 pCi/g Ra^{226} possible (low grade ores)
Decreased tailings maintenance
Ra^{226} isolation in mill
Increased by-product recovery possible (Th, RE's)
Increased uranium recovery
Economics for low grade ores seem attractive
Minimum solution to the tailings
Minimum neutralization required

Disadvantages

Not proven on complex ores (200 pCi/g Ra^{226})
Higher maintenance likely
Continuous integrated circuit not demonstrated on large scale

NEW H_2SO_4 FLOWSHEET

Consideration should be given to the possibility of modifying the present conventional H_2SO_4 flowsheet. The following is a proposed scheme to treat the uranium ores by a slightly modified and improved process. This is shown in Fig. 1.

The ore is slurried in water and is pressure-leached under oxidizing conditions to convert the sulphides to sulphuric acid which leaches the uranium. During the leach the radium is complexed so that it is rendered soluble [55]. Following leaching, the uranium is extracted form the slurry in a solvent-in-pulp process [56] and the solids passed to a wash thickener to recover any possible values. This wash solution is returned to leaching and a bleed stream of the original thickener overflow after uranium extraction is treated to control build-up of impurities before returning to the leaching stage. The gypsum from the treatment stage is blended with the thickner underflow solids and repulped in brine or dilute HCl to solubilize Ra^{226} which is subsequently isolated for disposal. The thickener underflow solids are water-washed before final discharge to tailings and the wash water returned to the chloride circuit.

The extracted uranium is stripped [57] and the uranium hydrolysed, enabling recycle of the strip reagent. Following filtration the product is dried to produce a high purity compound. The process has many advantages over any of the

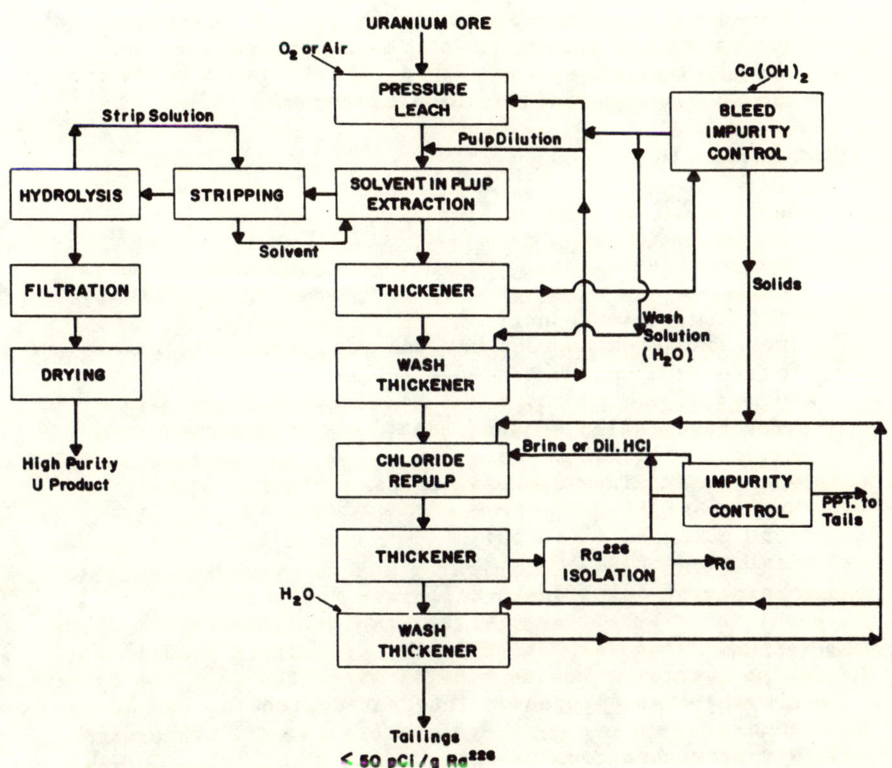

FIGURE 1 Future Uranium Flowsheet Incorporating SIP, Strip-Hydrolysis and Solution Recycle.

other flowsheets described. Table 5 lists these advantages. Although each of the segments of the flowsheet have been individually tested, an integrated circuit needs to be run in order to obtain mass and solution balances.

The extracted uranium is stripped [57] and the uranium hydrolysed, enabling recycle of the strip reagent. Following filtration the product is dried to produce a high purity compound. The process has many advantages over any of the other flowsheets described. Table 5 lists these advantages. Although each of the segments of the flowsheet have been individually tested, an integrated circuit needs to be run in order to obtain mass and solution balances.

TABLE 5 New H^2SO^4 flowsheet

Advantages

Smaller plant
Decreased L/S separation costs (operating & capital)
Higher purity product
Increased uranium recovery (decreased soluble losses)
Increased by-product recovery possible
Decreased energy costs
Decreased tailings management
Maximum solution recycle
Smaller tailings disposal areas
Minimum of lime requirements
  50 pCi/g Ra^{226} in tails
Isolation of Ra^{226}
Minimum of sulphide in tails

Disadvantages

Flowsheet not proven on integrated circuit (low grade ores)
Not proven on complex ores

DISCUSSION AND CONCLUSION

The main incentive for the development of new alternative processes is the concern for environmental factors. We must consider the two scenarios of radionuclides isolation in the mill as opposed to their disposal in the tailings. This is still an area of discussion. The flowsheets described here are based on the premise that a circuit has to have the radionuclides removed in the mill. Thus the radionuclides could be isolated, presumably for subsequent disposal. With the present practice of disposal in the tailings, the radionuclides gradually seep out of the impoundment area over time. The amounts or concentrations in the effluent have to be controlled by measured addition of $BaCl_2$. Although the tailings disposal areas are presently well designed and are continually being improved with respect to their stability, their operation and the need for perpetual maintenance can result in considerable and continuing costs.

REFERENCES

1. Brown, K.B. and Coleman, C.F., Solvent extraction in ore processing, Progress in Nuclear Energy; Series III, 2, pp. 3-34, PUb. Pergamore Press (1958).

2. Crouse, D.J. and Brown, K.B., ORNL-2720, Oak Ridge National Laboratory, Oak Ridge, Tennessee, 1959.

3. Tunley, T.H. and Nel, V.W., Proceedings of International Solvent Extraction Conference, ISEC' 74, Lyon, pp. 1519-1533, Pub. Soc. Chem. Ind., London, 1974.

4. Canadian Mining Journal, 81:6:75-149, June 1960.

5. Fisher, J.W. and Vivyurka, A.J., Ion exchange in the process industries, pp. 163-167, Pub. Soc. Chem. Ind., London, 1970.

6. Cross, R., J. South African Ind. Min. and Metall., pp. 196-201, Nov. 1968.

7. Clegg, J.W. and Foley, D.D., Uranium Ore Processing, Pub. Addison-Wesley, Reading, Mass., 1958.

8. Merritt, R.C., in The Extractive Metallurgy of Uranium, Pub. Colorado School of Mines Research Institute, 1971.

9. Ritcey, G.M. and Ashbrook, A.W., in Solvent Extraction - Principles and Applications to Process Metallurgy, Pub. Elsevier Scientific Publishing Company, Amsterdam, 1979.

10. Sialino, E., Mignot, C., Michel, P. and Vial, J., in Proceedings of International Solvent Extraction Conference, ISEC' 77, Toronto, pp. 592-599.

11. The Key Lake Uranium Process. Presented by Staff of Key Lake Mining Corp. and Sherritt Gordon Mines Ltd., at CIM Conference of Metallurgists, Halifax, Nova Scotia, August 1980.

12. Ritcey, G.M., Price, K.T., Wong, E.W. "Development of an Eluex process for the treatment of high chloride uranium eluates"; Presented at 1981 Annual Canadian Uranium Producers Metallurgical Committee Meeting, Saskatoon, May 1981.

13. Ritcey, G.M., Wong, E.W., Price, K.T. "Recovery of uranium from chloride solution resulting from HCl leaching of ores and concentrates"; Paper presented at International Solvent Extraction Conference, ISEC '83, Denver, 1983.

14. Ritcey, G.M. and Lucas, B.H., Proceedings of International Solvent Extraction Conference, ISEC '74, pp. 2437-2481. Pub. Soc. Chem. Ind., London, 1974.

15. Manfroy, W. and Gunkler, T., Proceedings of International Solvent Extraction Conference, ISEC '74, pp. 725-745.

16. Verhaege, M., Clause, G., Van Peteghem, A., VanAcker, P. "Diluent influence on the extraction of cobalt from chloride solutions with a tertiary amine"; Paper presente at 1982 AIME Annual Meeting, Dallas, February 1982.

17. Tremblay, R. and Bramwell, P., Trans. CIMM, 62:44:44-53, 1959.

18. Grimes, M.E. "The Technological Fugure of Canada's Uranium Industry"; paper presented at 86th Annual General Meeting, CIM, Ottawa, April 1984.

19. Bonney, C.F., Rowden, G.A. and McKee, D. "The combined mixer-settler: A new development in solvent extraction"; paper presented at Annual meeting AIME, paper A81-50, Chicago, Feb. 1981.

20. Roberts, J., Lyne-Smith, K., Collier, D.E. and McGee, J., Proceedings of International Solvent Extraction Conference, ISEC '77, Toronto, pp. 325-332.

21. Orjans, J.R., Notebaart, C.W., Godfrey, J.C., Hanson, C. and Slater, M.J., Proceedings of International Solvent Extraction Conference, ISEC '77, Toronto, pp. 340-346.

22. Barnea, E., Proceedings of International Solvent Extraction Conference, ISEC '77, Toronto, pp. 347-355.

23. Roberts, J. and McKee, J., Proceedings of International Solvent Extraction Conference, ISEC '80, Liege, paper 80-21.

24. Barnea, E. and Mizrahi, J., Proceedings of International Solvent Extraction Conference, ISEC '77, Toronto, pp. 374-384.

25. Ritcey, G.M. "Crud in Solvent Extraction Processing - A Review of Causes and Treatment"; J. Hydrometallurgy, 5, pp. 97-107, 1980.

26. Young, W. "Crud in Gulf Minerals Rabbit Lake Solvent Extraction Circuit"; Paper presented at AIME Annual Meeting, New Orleans, February 1979.

27. Kowalik, T. and Cantwell, T. "Crud Problems in Solvent Extraction and Strip Circuits of Conquista Uranium"; Paper presented at AIME Annual Meeting, New Orleans, February 1979.

28. Bakshani, N., Maurer, E.E., Allen, M.P. and Degenhard, A.L. "Crud at the Sohio Uranium Solvent Extraction Circuit; Paper presented at AIME Annual Meeting, New Orleans, February 1979.

29. Ritcey, G.M. "Crud in Uranium Solvent Extraction"; Proceedings of the 14th International Mineral Processing Congress, pp. VII-10.1-18, Pub. CIM, Toronto, October 1982.

30. Krebs Technical Literature on Solvent Extraction Systems, Krebs, Paris, France.

31. Castillon, P., Goodman, D. and Cheneau, G. "Industrial Experience with the Krebs Solvent Extraction System"; Paper presented at 1981 Conference of Metallurgists, Hamilton, August 1981.

32. Scuffham, J.B., The Chemical Engineer, pp. 328-330, July 1981.

33. Rowden, G.A., Dilley, M., Bonney, C.F. and Gillett, G.A., Proceedings of Hydrometallurgy 81, Manchester.

34. Germain, M. and Pluot, P., Proceedings of International Solvent Extraction Conference, ISEC '80, Liege, Paper 80-218.

35. Ritcey, G.M., Unpublished data.

36. Langdon, W.M., Sumpatchalit, T., Sampath, V. and Wasan, D.T., Proceedings of International Solvent Extraction Conference, ISEC '77, Toronto, pp. 413-418.

37. Prestridge, F., Johnson, B. and Longwell, R.L., Paper presented at 20th Conference of Metallurgists, CIM, Hamilton, Ontario, August 1981.

38. Wang, J. and Zhang, R., in Proceedings of Symposium "Hydrometallurgy"; Manchester, June 1981, sponsored by Society of Chemical Industry.

39. Lucas, B. and Ritcey, G.M., CIM Bulletin, Jan. 1975.

40. Hayashi, M., Nichols, I.L. and Huiatt, J.L. "Vanadium-Uranium Extraction from Wyoming Vanadiferous Silicates"; US Bureau of Mines Report of Investigations, R.I. 8822, 1983.

41. Ritcey, G.M., Lucas, B.H. and Ashbrook, H.W., in Proceedings of International Solvent Extraction Conference, ISEC '74, Lyons, France, pp. 943-968, Sept. 1974.

42. Ritcey, G.M. and Silver, M., CIM Bulletin, 75:846:134-143, Oct. 1982.

43. Skeaff, J.M. "Nitric Acid Leaching of Uranium Ore". CANMET Report MRP/MSL 75-165 (TR), Energy, Mines and Resources Canada, 1975.

44. Cable, R. (Mrs.) and Schulundt, H. "Some Experiments on the Extraction of Radium from American Pitchblende Ores by Chlorination"; Metallurgical and Chemical Engineering, XVIII:9:460-462, May 1918.

45. Suzuki, T., Izo, S., Tojama, S., Tomizawa, S. and Kodama, K. "Chlorination of Ningeyotge Uranium Ore by a Moving Bed Test Plant"; Tokyo Kogyo Shikensho Hokoku, 63, pp. 75-86, Feb. 1968.

46. Lapage, R. and Marriage, J.W. "Extraction of Uranium from Elliot Lake Ore by Vapour-Phase Chlorination"; Inst. Mining and Met., Trans. Sect. C, 82, pp. C101-C102, June 1973.

47. Gal, O.S. "Extraction of Uranium from Low-Grade Yugoslave Ores by Chlorination"; Proceedings of the International Conference on the Peaceful Uses of Atomic Energy, United Nations, Vol. 8, pp. 123-125, 1957.

48. Sawer, C.W. and Handley, R.W. "Process of Extracting Uranium and Radium from Ores"; U.S. Patent 2,894,804, 1959.

49. Ewing, R.A., Foley, D.D., Beause, A.E. and Filbert, R.B. "Nonaqueous Extraction Methods for Western Uranium Ores"; Battelle Memorial Institute, Report No. BMI-278, Oct. 1954.

50. Skeaff, J.M. and Laliberte, J.J. "Continuous High Temperature Chlorination of Uranium Ores"; Presented at AIME Meeting, Portland, Oregon, Division Report ERP/MSL 80-48 (OP), CANMET, Energy, Mines and Resources Canada, Aug. 1980.

51. Haque, K.E., Lucas, B.H. and Ritcey, G.M., CIM Bulletin, pp. 144-147, 1980.

52. Ritcey, G.M. and Wong, E.W. "Solvent Extraction from a Chloride Leach Liquor"; Presented at ISEC '83, Denver, Aug. 1983.

53. Demopoulos, G. and Ritcey, G.M., CANMET, Energy, Mines and Resources Canada, Patent Pending.

54. Ritcey, G.M., Slater, M.J. and Lucas, B.H., Proceedings of International Symposium on Hydrometallurgy, AIME Annual Meeting, Chicago, Feb. 1973, pp. 419-474, Pub. AIME' 1973.

55. Ritcey, G.M. and Weatherell, C., CANMET, Energy, Mines and Resources Canada, Patent pending.

Uranium Enrichment by Solvent Extraction

J.M. Lerat and C. Lorrain, Commissariat à l'Energie Atomique (IRDI/DESICP) France

INTRODUCTION

For a long time, the uranium enrichment processes by chemical exchange were considered by the experts [1] as not commercially viable for two main reasons :

- a low elementary isotopic effect
- slow kinetics of the chemical exchange reactions

In 1968, however, these difficulties were simultaneously overcome by the CEA who discovered, in a liquid liquid system, an important isotopic effect accompanying an instantaneous chemistry [2], and the ASAHI Japanese company who developped reaction catalyzers and a particulary efficient resin.

Since this progress, the chemical processes seemed so attractive that they became the second generation solution in the French strategy of the development of enrichment techniques (after the Gaseous Diffusion technique).

The present report presents the current development programme of the French chemical process (named CHEMEX) and describes its latest achievements.

.../..

PRINCIPLE AND MAIN THEORETICAL FEATURES OF THE PROCESS

Elementary effect [3]

The process is based on an isotopic effect observed during an exchange between two uranium ionic species U_A and U_B, reacting according to the following relation :

$$235\ U_A + 238\ U_B \rightleftharpoons 238\ U_A + 235 U_B$$

The reaction equilibrium constant characterizes the isotopic separation factor α.

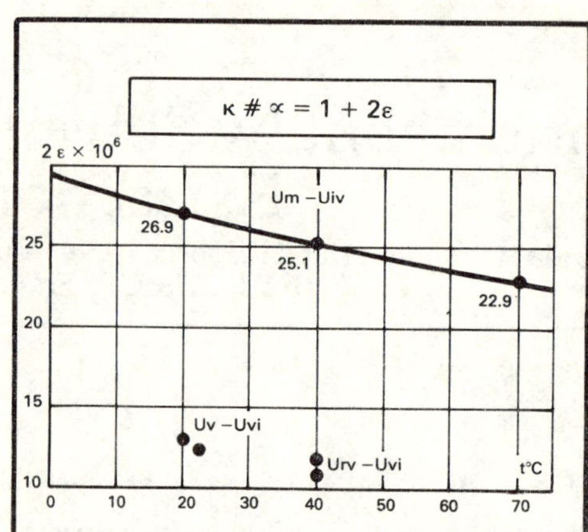

FIGURE 1 CHEMEX separative effect

The originality of the French process lies in the choice of the interacting species U_A and U_B : trivalent (U_{III}) and tetravalent (U_{IV}) uranium..

The elementary effect reaches 1,025. The corresponding enrichment coefficient is about the double of the ones encountered in the $U_{IV} - U_{VI}$ or $U_V - U_{VI}$ exchanges. This coefficient increases as the temperature decreases (figure 1).

Phase choice and exchange kinetics [3]

Enriched and depleted fractions resulting from the reaction must be separated before entering respectively into new reactions. This is done by means of two non miscible phases which capture selectively the uranium compounds. In the case of the CHEMEX process two non miscible liquid phases are counter currently contacted : an aqueous hydrochloric phase retaining the trivalent uranium and an organophosphorous phase retaining the UCl_4 compound. After equilibrium the lighter organic phase is enriched in U_{235} isotope.

The U_{III}-U_{IV} kinetics is so quick that it was never possible to observe a transient equilibrium. Thus the overall kinetics is controlled by the interfacial exchange kinetics

.../..

Cascading and process flowsheet [4]

In spite of a relatively important isotopic effect, the enrichment needed for the nuclear reactor fuel requires the cascading of several thousands of separation stages. Individual or differential stage contactors can be used.

To operate a solvent extraction enrichment cascade both ends of the isotopic exchange section are equipped with head and tails refluxes aimed at :

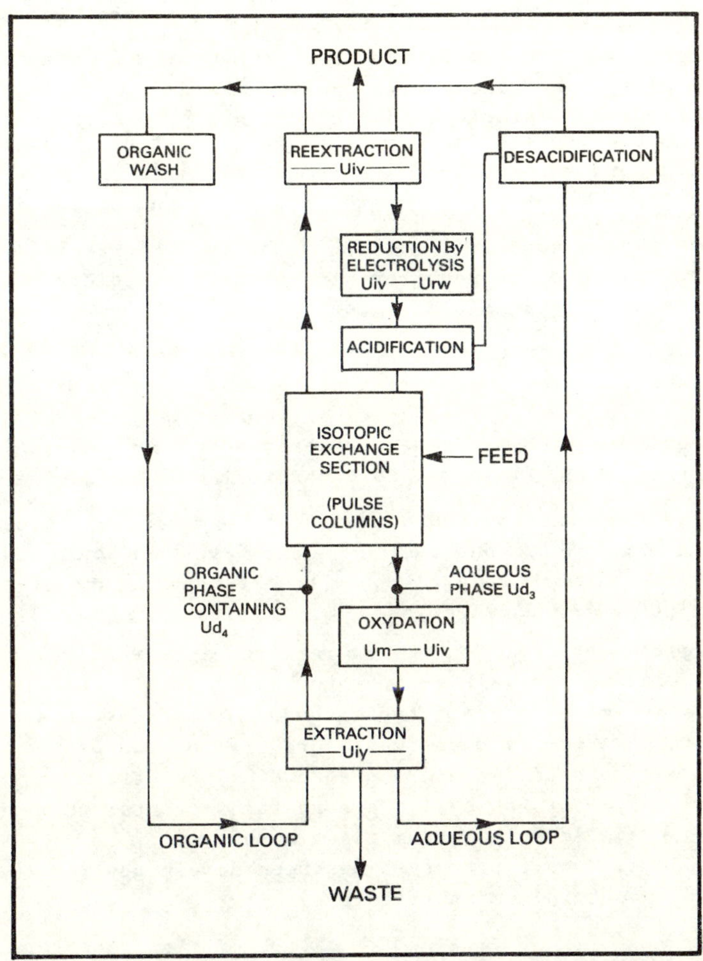

FIGURE 2 CHEMEX simplified flowsheet

- the realization of the uranium valence changes.
- the extraction of the uranium from the aqueous phase (tails reflux) and its stripping from the organic phase (head reflux).
- the closing of the three process loops : the uranium loop, the aqueous and the organic loops (figure 2).

CHEMEX TECHNOLOGICAL FEATURES AND REQUIREMENTS [5]

For CHEMEX industrial units most of the equipment items can be provided by the conventional chemical industry. However their availability and their performance had to be evaluated in relation to the specific CHEMEX operating conditions. To do this

a large development programme has been achieved covering more particulary the following technology fields : materials of construction, isotopic exchange contactors, electrolyzers, process equipment (pumps, heat exchangers, tanks, piping fittings and accessories, instrumentation).
The choice of materials depends primarily on the chemical

nature of the handled liquors. In the isotopic section fluorocarbon polymers are required. In the other sections more conventional materials are accepted.

The CHEMEX isotopic exchange is performed by contacting the reacting phases in a pulsed column, operating in controlled, well defined hydraulic conditions, made of plastics, and equipped with membranes to separate the process solutions from the moving fluid of the mechanical pulsing device.

For the construction and operation of numerous plastics columns, reliable construction codes have been succesfully formulated, that are applicable to the manufacture of contactors 1600 mm in diameter and 30 m in height (figure 3).

The aqueous phase reduction is realized in a vertical filter press electrolyzer. This apparatus has a modular construction. Anodes are made of graphite, cathodes of mercury. A separator kept in position by plastic spacers, is mounted between each anode and cathode. Anodic and cathodic liquids are introduced at the bottom of the cell. Mercury falls counter currently and is recycled by a bottom plastic pump.
A 50,000 A, 7 cell prototype has been developped for the CHEMEX industrial units (figure 4).

.../..

COLUMN DIAMETER (m)	COLUMN HEIGHT (m)	CONSTRUCTION MATERIALS (SETTLERS AND BODIES)
0.100 0.150	1.8	PVF_2, GLASS REINFORCED POLYESTER
0.225	1.5	FEP. PTFE OR PVF_2 LINED STEEL GLASS REINFORCED POLYESTER
0.350	21.1	CARBON FIBER REINFORCED PVF_2
0.400	5	FEP AND PTFE LINED STEEL
0.600	1.8	FEP, PTFE, GLASS REINFORCED POLYESTER
1.200	1.8	FEP LINED STEEL

FIGURE 3 Experience in the construction of plastics pulsed columns

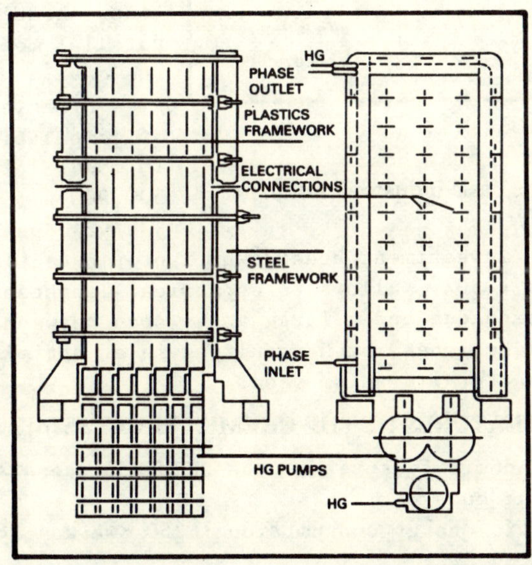

FIGURE 4 7 cell electrolyser

INDUSTRIAL AND COMMERCIAL FEATURES OF THE CHEMEX PLANTS

The choice of the pulsed column contactors and the necessity of large refluxes lead to a very particular plant structure.

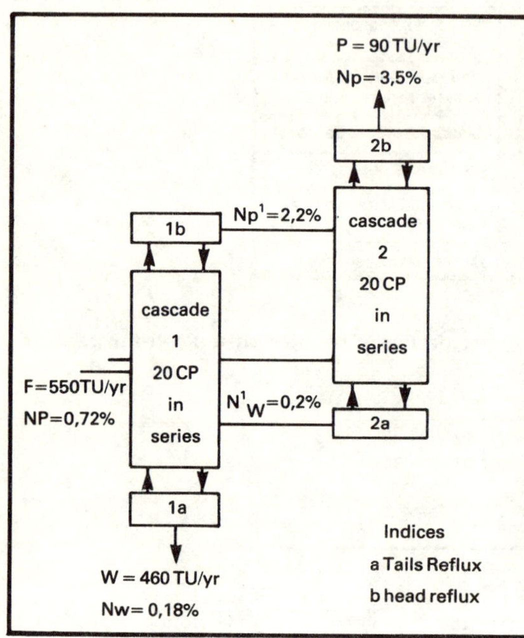

FIGURE 5 The basic industrial module (500,000 SWU/yr)

The basic element is the 1200 mm diameter, 30 m. high, pulsed column, having an isotopic separation power of about 15,000 SWU/yr. The CHEMEX projected facilities are made out of twin square cascade modules as shown in figure 5. In the case of a 0,18% waste assay, 20 identical basic pulsed columns are connected in series in each cascade. The module overall capacity amounts to 500,00 SWU/yr.
An industrial facility may include, according to the customer's production schedule, several parallel modules.
The developped technologies offer a good level of competition for I to 10 module plants, be for 500,000 to 5.000,000 SWU/yr capacities.

The specific investiment breakdown (isotopic section : 25% and refluxes : 50%) allows an economical lenghtening of the enrichment cascades and a large assay span between the depleted and enriched ones, both resulting in an optimized utilization of the natural uranium feed.

ECONOMIC FEATURES OF THE CHEMEX PROCESS [6]

The main economic features of the CHEMEX process are presently the following ones :

- the specific energy consumption : 250 Kwh and 380 thermies by delivered SWU.

.../..

- the specific uranium inventory : 1,07 kg U by delivered SWU

These two intrinsic values are independant of the size and economic environment for the project.

- the specific investment : 3535 FF by installed SWU (1984 cost) including the construction and commissioning expenses, so as the interests during construction.
- the operation and maintenance costs : 95 FF by delivered SWU (1984 cost).

These values are directly related to the plant size (6 modules cost) and the economic environment (realization in France).

The result of these economic assessments is a SWU unit cost of about 640 FF (1984 cost).

PRESENT ACHIEVEMENTS AND FUTURE PERSPECTIVES OF THE CHEMEX

Present achievements

The results of fifteen years of R and D may be summarized by the present attainments balance :

- Choice and definition of optimized phases.
- Control of the stability of the phases.
- Steady increase of the isotopic and interface exchange efficiencies quantified by performance indicators (figures 6 and 7).
- Successful operation of U_{IV}-U_{III} electrolyzers resulting in a regular decrease of the specific electricity consumption (figure 8).
- Choice of a plant structure perfectly fitted to the used contactors.
- Realization of several plant preprojects in collaboration with the KREBS and TECHNIP companies followed by a successful comparison of the project cost with the American Department of Energy estimations.

The resulting cost of the CHEMEX SWU is about the same as the SWU cost of an amortized Gaseous Diffusion plant using the French technique.

Future perspectives

The CHEMEX industrial feasibility being already assessed, the present development programme is primarily aimed at the increasing of the project economic competitiveness and performances. Two achievements are in progress :

- A 1/10 th scale process pilot plant (named PP 35) to demonstrate the good operation capability of the CHEMEX process.

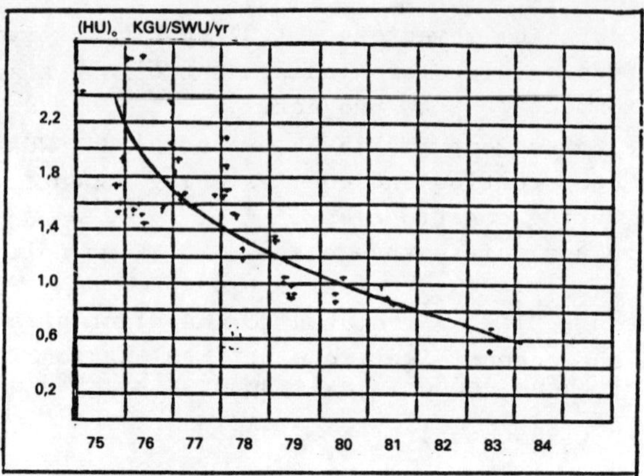

FIGURE 6 Specific uranium inventory

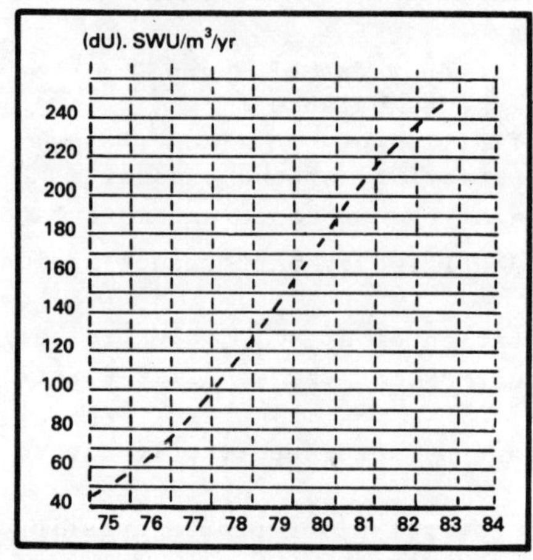

FIGURE 7 Volumetric separation capacity

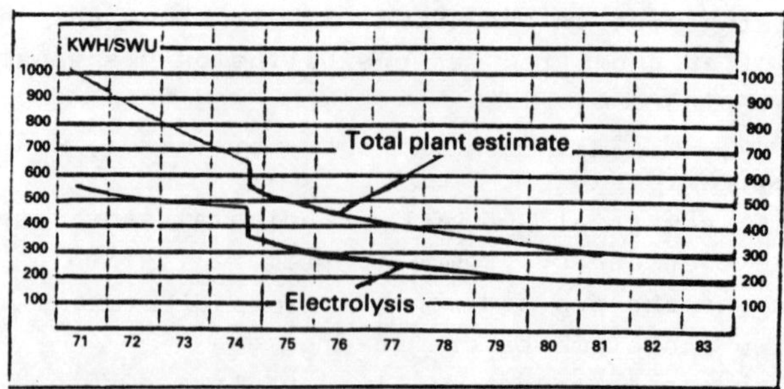

FIGURE 8 CHEMEX electricity consumption

This unit has just been commissioned and isotopic tests are scheduled in 1985.

- A process consolidation programme in two pilot units : the one being dedicated to the reflux rationalization (named PL 8) and the other to the optimization of a cascade composed of the same stage number as the commercial plants.

CONCLUSION

Inspite of the depressed state of the enrichment needs the CHEMEX process keeps offering great advantages even though the priority of the French development programme is devoted, since 1984, to the third generation advanced techniques (laser and cyclotronic resonance), namely : an excellent economy at small scale, the use of conventional techniques and a great operation flexibility.

References

(1) BENEDICT M. and al., report on uranium isotope separation review committee - ORO 694 USAEC report, June 1982.

(2) DELVALLE P., French patent N° 74.29111, August 8 th,1974 Procédé et dispositif d'échange isotopique. US Patent 4,012,480, March 15 th, 1977.

(3) COATES J.H., FREJACQUES C. and LERAT JM. Uranium isotopic separation through chemical exchange,ANS Winter meeting, San Francisco USA, December 1st, 1977.

(4) COATES J.H., Present state of the French chemical enrichment process (CHEMEX), American Institute of chemical engineers, (1982) Winter National Meeting, ORLANDO (USA)

(5) LERAT JM. LORRAIN Ch. SAUZAY G. and VENOT P. Improving the commercial competitiveness of CHEMEX, Nuclear Engineering International, January 1984, p. 25-26.

(6) BERNAUD C. COATES J.H. GELEE M. LERAT JM. and PLURIEN P. Etat d'avancement du procédé d'enrichissement de l'uranium par échange chimique, International conference on nuclear power experience, Vienna, 13-17 September 1982 IAEA - CN - 42/168.

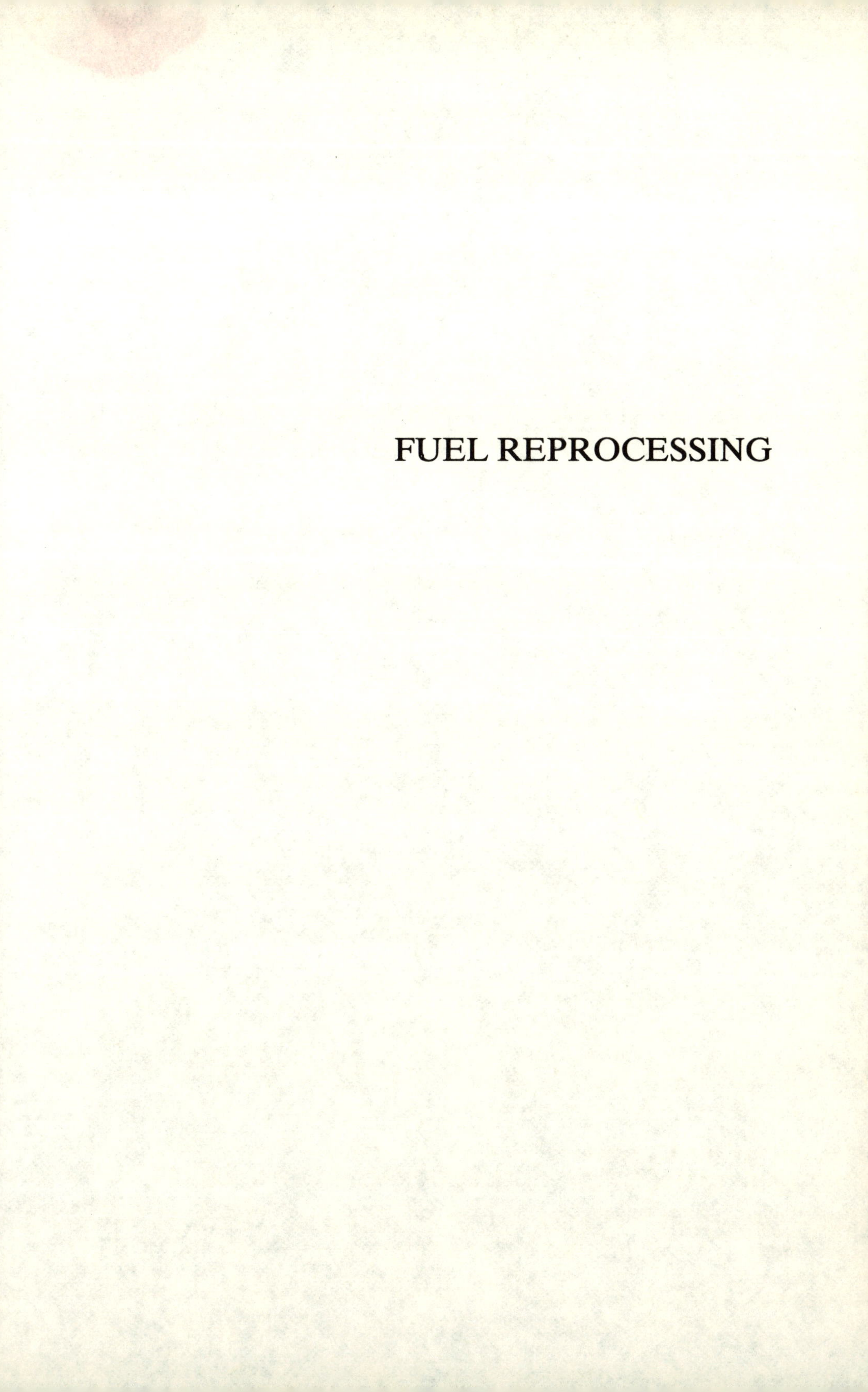

FUEL REPROCESSING

Reprocessing of Fast Breeder Reactor Fuels in France

Louis Patarin, Coordonnateur Pour Le Retraitement Des Combustibles Nucléaires a l'Institut de Recherche Technologique et de Developpement Industriel (C.E.A.)

INTRODUCTION

The reprocessing of breeder reactor fuels is a direct technical descendant of the reprocessing of thermal reactor fuels which was developped first. The process used is in both cases the PUREX process, which consists in dissolution by nitric acid followed by selective extraction using TBP. In France, the application of this technique to breeder reactor fuels greatly benefited from the scientific and industrial experience initially acquired with metallic fuels of the MAGNOX type and then with oxide fuels of the LWR type.

Another feature of French advances in this sector is the constant monitoring of the programs for, on the one hand, fast breeder reactors and, on the other, their fuel cycles, including reprocessing. Figure 1 summarizes French developments in reprocessing breeder reactor fuels with reference to the parallel development of the reprocessing of thermal reactor fuels and fast breeder reactor.

FRENCH ACHIEVEMENTS

The experimental reactor RAPSODIE was associated with the AT1 reprocessing pilot facility on the La Hague site, which entered service in 1969, two years after the reactor, and with a small capacity (1 kg/day). Until 1979, when it was definitively shut down, this shop processed one tonne of heavy metals from mixed oxides irradiated to a burn-up of up to 120,000 MWd/t and sometimes only slightly cooled ($1\frac{1}{2}$ months for some fuels). It thus closed the RAPSODIE cycle several times and at the end of

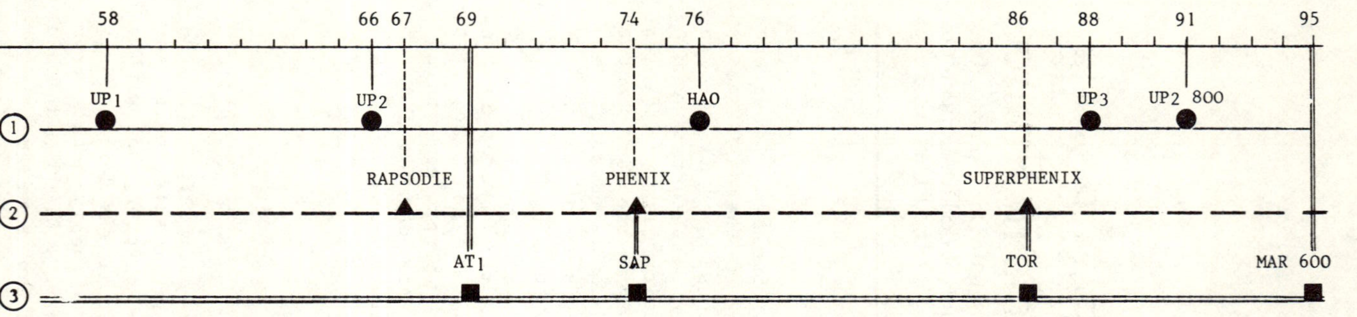

(1) THERMAL REACTOR FUEL REPROCESSING.

(2) FAST BREEDER REACTOR.

(3) FAST BREEDER REACTOR FUEL REPROCESSING.

UP_1	MARCOULE PLANT (MAGNOX).
UP_2	LA HAGUE PLANT (MAGNOX).
HAO	LA HAGUE OXIDE HEAD END.
UP_3	LA HAGUE OXIDE PLANT (800 t/y) 1st UNIT.
UP_2 800	LA HAGUE OXIDE PLANT (800 t/y) 2d UNIT.
AT_1	LA HAGUE FBR REPROCESSING PILOT PLANT.
SAP	MARCOULE FBR REPROCESSING PILOT PLANT.
TOR	SAP EXTENSION.
MAR 600	MARCOULE FBR REPROCESSING PLANT (50 t/y).
RAPSODIE	EXPERIMENTAL BREEDER.
PHENIX	DEMO BREEDER (250 MWe).
SUPERPHENIX	INDUSTRIAL BREEDER (1200 MWe).

FIGURE 1

		initial Pu / U + Pu	Years	t
AT_1 LA HAGUE PILOT PLANT	RAPSODIE	25 - 30 %	69 - 79	0.91
	PHENIX	18 %	78 - 79	0.18
SAP MARCOULE PILOT PLANT	RAPSODIE	30 %	75	0.05
	KNK 1	ENRICHED U	75 - 76	1.65
	PHENIX	ENRICHED U	77 - 78	2.30
	PHENIX 1	18 %	79 and 83	0.380
	PHENIX 2	25 %	79 - 83	6.36
HAO-UP_2 LA HAGUE	PHENIX 1	18 %	79 - 84	9.91
			TOTAL	21.74 t

FIGURE 2

its life, allowed experimentation with PHENIX fuel reprocessing.

PHENIX, however, required a new pilot facility, the SAP at Marcoule, in association with HAO oxide head end of the UP2 plant at La Hague, which uses dilution in the MAGNOX fuels. After processing various fuels, in particular from RAPSODIE, KNK1 and the first enriched uranium core of PHENIX, the SAP facility reserved its processing activities for the PHENIX fuel which was the richest in plutonium (25%), and from 1979 to 1983, processed a total quantity of 6.3 tonnes. For its part, the HAO oxide head end processed almost 10 tonnes of PHENIX fuel (18% of plutonium), the last processing period taking place at the end of 1984.

Figure 2 shows the production figures with the various French resources.

PROJECTS

TOR - The Marcoule pilot facility is undergoing renovation and extension to become the future TOR plant. The capacity will thus be extended to 5 tonnes of heavy metal per year. The research potential will also be increased, thanks to specific cells and equipment which will make up the TOR "advanced technology units".

The new pilot shop comprises three sections. Installed in a new building, the largest section, TOR.1, comprises twenty or so cells containing the process head end equipment. This building houses the receipt and storage facilities for the irradiated fuels, the mechanical processing, chemical dissolution and solution clarification units, as well as the waste storage and packaging facilities. The main head operations may be performed either using conventional equipment, or with the "advanced technology units" which can operate in parallel with the main line. TOR.1 is equipped with a high degree of remote control and maintenance equipment.

The other sections, TOR.2 and TOR.3, are located in the premises of the old pilot shop. TOR.2 comprises storage facilities for finished products in the form of purified nitrate (U and Pu) and TOR.3 comprises a new pulsed column extraction first cycle with the possibility of partition, as well as a new evaporator for fission products solution.

The TOR will enter service at the beginning of 1986.

MAR.600 - Between 1979 and 1982, the CEA Group studied a project for a large plant capable of reprocessing the fuels from 6 1500 MWe fast breeder reactors (PURR project), since a plant of this capacity corresponded to the industrial development programme envisaged at that time. This project showed that the technical bases existed for a plant of this size and gave a clear indication as to the economic aspects.

After the construction of SUPERPHENIX (1200 MWe), present fast neutron reactor strategy foresees a fairly long intermediate stage on a fully European scale which will precede true industrialisation on a large scale. This intermediate stage could comprise two 1500 MWe reactors to enter service before the year 2000. This is why the new project for a French plant only had a capacity of 50 tonnes/year, which would cover the needs of SUPERPHENIX and the two subsequent large reactors. The European scale was broadened in 1984 with the entry of the United Kingdom and a plan is afoot for this project to become Franco-British. The following description only refers to the French study planned for the Marcoule site (project MAR.600) and entrusted to SGN by COGEMA.

The characteristics of MAR.600 are as follows:

- optimum use of the existing infrastructures and general services of the Marcoule centre,

- a single equipment line,

- fuels cooled for three years with a maximum burn-up of 125,000 MWd/t,

- study schedule compatible with a decision for construction in 1986,

- UP3 technology (new La Hague plant) when applicable, or specific technology of the R and D programme described in paragraph 3.

EXPERIENCE ACQUIRED AND R AND D

Although the reprocessing technology for fast breeder reactor fuels is to a large extent common to that of PWR reactor fuels, additional experience is needed for industrial implementation. The specific characteristics of fast breeder fuels which need in particular to be taken into account are:

- the assembly structure which comprises small-diameter fuel pins in stainless steel cladding in a thick case, thus making bulk shearing impossible,

- the mixed oxide composition with a high concentration of plutonium (10 to 20%), which complicates the criticality problems and makes dissolution more difficult,

- the level of fission products, which considerably increases the thermal power at all stages at which they appear.

The results obtained in the various French plants (paragraph 1) on significant quantities, are already providing extremely valuable lessons for the future. In particular, the pilot facility allowed considerable experience to be gained, the major points of which are as follows:

- trapping of iodine released upon dissolution can be extremely efficient,

- ruthenium is the major contaminant of cladding debris (solid waste),

- the hydraulic behaviour of pulsed extraction columns is satisfactory under diverse conditions,

- decontamination, measured for each extraction cycle, is excellent since the levels of uranium and plutonium meet the standard values as early as the end of the second cycle,

- mixer-settler electrolysis partition or partition with hydroxylamine nitrate plus a valence 4 uranium filler is perfectly satisfactory,

- finally, the plutonium balance is well controlled since the total divergence over five cycles is 0.48%.

The pilot shop has also allowed innovative experiments, such as the processing of the solvent by hydrazine carbonate or trapping of iodine on solid absorbants, to be performed.

In addition to this collection of operating results, the laboratories of the CEA are running studies and tests which are today based around the MAR.600 project. This extremely important research programme (80 million francs in 1985) can be summarized by looking at the main techniques envisaged for use in the plant.

Mechanical Processing

The first phase consisting of the dismantling of the fuels will use H.F. electrical cutting to remove the upper end-piece. In the second phase, the case will be opened by TIG cracking by embrittlement with a special metal electrode. This latter method has already been successively tried with four irradiated PHENIX fuels.

The fuel pins will then be extracted from the bundle one after the other and will pass through a machine which will remove the spacing wire using a "rotating arc". The mechanical processing will be completed by chopping the fuel pins into 30 mm sections in a rotating shearing machine, the blade of which may be remotely replaced.

The case will also be broken down into shorter pieces for waste disposal by means of cracking by embrittlement.

Chemical Processing

Dissolution presents both chemical and equipment problems. In hot cells, the CEA defined optimum conditions for oxide etching by using nitric acid and finalized a high-performance dissolution process to supplement the main dissolution process if necessary.

The cladding sections will be dissolved in a continuous helicoidal circulation apparatus. This apparatus has already been tested in a non-active environment and will first of all be operationally tested in a TOR "advanced technology unit".

The solution may be clarified in a centrifuge of the type developed for UP3 in La Hague.

The extraction cycles will use "salt-free processes". Thus separation using hydroxylamine nitrate and processing of the solvent with hydrazine carbonate will be developped. The neptumium content will be oriented towards the plutonium flow for recycling in the reactor. The second plutonium cycle will be a concentrating cycle and a reflux will provide the necessary flexibility. The second uranium cycle will also be a reflux cycle. This flowsheet will be proven in the TOR facility. As far as the equipment is concerned, CEA proposes pulsed columns with disk-crown packing. Finally, in order to reduce alpha emitters release, there will be "alpha traps" which will allow reextraction of plutonium traces from the solvent prior to its processing.

Production of Pu02

MAR.600 will use new techniques for oxalic precipitation and oxalate calcination: rotating bowl precipitator, flat filter, double-screw calcinator.

The rectification of plutonium outside standardized values may be done using a new extremely efficient dissolution process.

Processing of Fission Products

The CEA will adapt its vitrification process to the fission products solutions produced in MAR.600. It will also attempt to improve its technology by creating a high temperature fusion furnace operating with direct electric induction.

Waste Processing

All plant waste will be processed and packaged: for example, the iodine will be trapped on solid silver supports and clad waste will be melted by direct induction.

Remote Control and Maintenance

The general architecture of the plant will be compact and with no duplication of lines. The main workshops will be equipped with modern remote control resources which are more reliable and more mobile than conventional mechanical master-slave equipment. In particular, an electronically controlled servo-manipulator designed by CEA and already widely operational will be used (MA 23 M manufactured by La Calhène).

Instrumentation

MAR.600 will be equipped with in-line analysis devices and will for example use fibre optics spectrophotometry.

CONCLUSION

Thanks to the wealth of experience already acquired in its laboratories and plants, France can be confident of technical success in its future installations, in particular in the field of fast breeder reactor fuel reprocessing. A constant R and D effort is resulting in constant progress towards a level of economic success which is the future challenge for this type of industrial activity. European cooperation, sought after and actively undertaken by those countries interested in the fast breeder reactors, will confirm this hope before the year 2000 if it is based around concrete projects.

Fast Reactor Fuel Reprocessing Experience in Germany

K.H. Ebert, Nuclear Research Centre Karlsruhe

INTRODUCTION

In the Fed. Rep. of Germany fast breeder energy production was considered a prospective development since the early sixties /1/. An experimental reactor (KNK-II) has been in operation in Karlsruhe since 1979 and works without any serious problems, some of the fuel elements having achieved a burnup of more than 100000 MWd/t equivalent. The 300 MWe prototype breeder reactor in Kalkar (SNR-300), a joint German-Belgium-Netherland project, is expected to go into operation in 1986. Design studies for a 1300 MWe breeder power reactor (SNR-2) are in progress.

Work on reprocessing of breeder fuel in the Fed. Rep. of Germany started about 25 years ago, but was limited to R&D. At the present time Germany is not planning the construction of a commmercial breeder reprocessing plant, but there is considerable interest in participating in an European plant designed, constructed and operated jointly by France, Great Britain and the Fed. Rep. of Germany.

R&D PROGRAM

Most of the R&D work on fast reactor reprocessing in the Fed. Rep. of Germany is done at the Nuclear Research Centre, Karlsruhe (KfK) in the 'Institut für Heisse Chemie' (IHCh), which was founded in 1959 for that purpose /2/. The IHCh R&D program was concentrated on the three topics: (1) process selection, (2) waste minimization and (3) process optimation.

In the early years, a number of different processes were studied until the decision was made, around 1965, that the Purex process should be favoured for a commerical breeder reprocessing plant /3/. It was recognized that there were a number of weak points in the Purex process, which needed intensive R&D work to adapt it to breeder conditions, which are characterized chiefly by a higher radioactivity, a 10 times higher plutonium/uranium ratio, and difficulties with quantitative dissolution of the fuel. There was also the rather critical disadvantage that this process results in at least 3 different waste streams of varying radioactivity, which have to be processed separately. The reason for this is that different chemicals are used in the redox process and solvent wash operations, which create a relatively high waste volume in the medium active range. One of the early projects of the IHCh was to develop salt-free processes either by electrochemistry or by chemicals, which can be decomposed into gases which can be released to the atmosphere through the plant off-gas system. The long term research program covers the following projects: (1) dissolution studies and characterization of residues and precipitates, (2) flow sheets studies on the extraction and purification cycles, (3) electrochemical reduction and oxidation of plutonium under Purex conditions, (4) behaviour of technetium, ruthenium and transuranium elements in the Purex process, (5) development of an integrated off-gas system including the retention of krypton, (6) development of new purification cycles, (7) studies of the mechanisms and kinetics of the radiation reactions on the solvent constituents.

No less important is work dealing with criticality considerations, improved availability and adjustment to optimal operation of the plant, and minimization and control of the plant inventory. For those purposes research projects are in progress on (1) development of inline instruments for detecting plutonium, uranium, fission products, and nitric acid, (2) designing new equipment (dissolver, extractors, evaporators etc.) with long life-times, (3) developing mathematical models for simulating optimal- and maloperations and supporting the layouts of equipment and instrumentation, and (4) development of remote techniques including robotics and adapting conventional equipment for these techniques, especially for such instruments for which a limited operational life has to be assumed.

FLOW SHEET STUDIES

The investigations on the applicability and optimization of the Purex process in the earlier stages of the work were carried out with LWR-UO_2 /4,5/ and MOX (recycle) fuel and more recently with FBR-MOX fuel /6,7,8/. The equipment available in IHCh for these studies is:

- LABEX (laboratory scale extraction facility) in operation since 1966, mixer settlers, throughput: 0.5 kg/d, shielded suitable for U, Pu, Np, Tc,
- MILLI (fully shielded Purex mini plant), 1971, mixer settlers, 1 kg/d, any original fuel.
- UTE (column test facility for uranium) 1974, 200 kg U/d.
- PUTE (electrochemical column test facility), 1981, 100 kg U+Pu/d, aplpha shielded
- MINKA (miniature pulsed column test facility) 1984, 20 kg U+Pu/d, alpha shielded

Early investigations in LABEX were focussed on the extraction and U-Pu separation of breeder fuel /9/. Using feeds of up to 30% Pu in the heavy metal and extractants of up to 30% TBP, the extraction efficiency, the Pu losses in the raffinates and the organic phases, as well as accumulations during the extraction were studied. The Pu-U separation by U-IV was optimized in LABEX, and the U-IV excess minimized to the 6 fold stochiometric value; these results were verified in MILLI.

The instabilities of the U-IV reduction and the high excess recycle rates of U-IV provided the motivation for developing the electrochemical reduction of Pu inside the extractor /10,11/. The successful construction of an electrochemical mixer settler for Pu-U-partitioning and Pu back-extraction in LABEX was followed by introducing this new method into the 3rd cycle of MILLI, and in 1980 the technical scale was reached by introducing an electrochemical mixer settler into the 2nd Pu cycle of the WAK plant /35/. Further advantages of the electrochemical extraction are higher decontamination factors for Pu and more stable operational conditions. Analogous results were obtained in the electrochemical pulsed column (ELKE) which is installed in the plutonium test facility (PUTE) /12/. D.F's of up to $9 \cdot 10^5$ for Pu were obtained at Pu concentrations of 40 g/l in the plutonium purification cycle. For the Pu/U separation these D.F's were up to $1.8 \cdot 10^3$ obtained under the rather unfavourable conditions of 30% flooding. Electrochemical processes were also developed for the oxidation of Pu-III /13/; and those ROXI-cells were successfully tested in LABEX and MILLI. The equipment was scaled up to the pilot plant level and successfully demonstrated in the WAK plant.

In an effort to reduce the medium active waste volume, hydrazine carbonate was used as a washing agent instead of sodium carbonate /14/. The washing effect of both agents is approximately the same but excess hydrazine can easily be oxidized electrochemically to gaseous nitrogen without producing additional waste.

Another electrochemical process application is the decomposition of excess nitric acid, and oxalic acid from Pu precipitation mother-liquors /13,15/.

In extensive investigations, typical separation and critical data for Pu-III and -IV, U-IV and -VI, Np-IV, -V and -VI, /16,17/, Tc-IV and -VII /18,19/, nitric acid and hydrazine were determined, and were verified in LABEX and MILLI. Considerable effort was employed in recent years to develop a mathematical model for simulating the PUREX process in mixer settlers and pulsed columns, including the electrochemistry. For this work a considerable number of partition data /20,21/ was needed as well as data on the transfer kinetics of certain species /22/ and the characteristics of the back-mixing behaviour of the columns. All available data from our own work as well as from others were included in the model. Now the model is so well developed that valuable and valid information can be obtained on the performance of columns and mixer settlers in the laboratory and on the technical scale /23,24/. The simulation of mal-functions and the consequences of mal-operations from unsuitable flow rates or incorrect feed adjustments can quickly be calculated. The model sumulations proved to be extremely helpful in defining experiments and evaluating results in PUTE and MILLI. For studying non-regular operational conditions including the problem of Pu accumulation, the column test facility MINKA was set up which went into operation in 1984.

Within the last 14 years, roughly 150 kg fuel (about 110 kg LWR-fuel, av. burnup 45000 MWd/t, cooling time 1-7 y /5,14,/and about 40 kg FBR-fuel, 60000-100000 MWd/t, 0.8-2y) /6,7,25/ were reprocessed in MILLI in 23 campaigns. The breeder fuel comes mainly from KNK-II: fuel element disassembly and pin shearing was carried out in KfK's Hot Cell Facilities /45/. It was demonstrated that with the Purex process all commercial fuel can be reprocessed, and the required product specifications can be obtained. In most cases these specifications could be obtained with a 2 cycle process.

The kinetics of the radiolytic reactions of TBP to form HDBP and other decomposition products were studied as well as their effects on the extraction behaviour in mixer settlers and pulsed columns. The following topics were investigated in greater detail: (1) decomposition of TBP and dodecane by hydrolysis and radiolysis /26/ (2) identification of complex compounds which reduce the usefulness of the solvent /27/, (3) effects on the distribution factors of U, Pu, Zr, Nb, Ru /28/ (4) variation of the behaviour of the 2 phase-system relative to the separation of emulsions or the wetting of the sieve plates, (5) variations in the extraction kinetics and (6) formation of crud.

A considerable amount of work has been devoted to develop new processes for the purification of both aqueous and organic streams to improve recycling. Ion exchange resins were used to purify the aqueous feed from HDBP and other radiolysis products /29,30/. This process is considered superior to the steam wash for several reasons.

It is considered important to develop new methods and automatic instrumentation for process analysis. This is a rather large field, related to the many different substances and the huge range of concentrations, which govern the performance of the Purex process /31/. Precise analytical data are necessary for the attempt to identify optimal process conditions /32,33/. For maintaining optimal conditions rapid analytical information is needed, which is best obtained from inline instrumentation. An alpha monitor (rotating drum) has been developed and successfully tested in PUTE /34/; it was recently installed in the WAK plant for technical demonstration purpose. More development work on other inline instruments, e.g. energy dispersed X-ray fluorescence, density and electrical conductivity, is in progress.

DEVELOPMENT OF EQUIPMENT

It is an important part of the IHCh work program to develop new equipment up to pilot plant scale and to demonstrate its satisfactory performance in plant operation. Whenever possible attempts are made to install new equipment in the WAK pilot plant. After rigorous testing on the laboratory scale, designing and scaling up is done by a joint group of IHCh and WAK personnel. The new equipment is then constructed, and test runs are carried out in the IHCh, again jointly by personnel from both organizations. Due to the rather good cooperation and to mutual interest a relatively large number of new types of equipment could be put into the WAK plant for technical demonstration. This procedure, of course, is restricted to equipment adaptable to the WAK plant. As the WAK is equipped with mixer settlers, it cannot be used for testing extraction columns. Therefore a numer of column test rigs (UTE, PUTE, MINKA) had to be set up within the IHCh itself.

Development of electrochemical mixer settlers started with the development of equipment for MILLI and LABEX. A pilot plant size electrochemical mixer settler (2B EMMA) was constructed to replace the original mixer settler of the 2nd cycle in the WAK plant /35/. At the same time a pilot plant size flow-through cell for electrochemically oxidizing Pu-III (ROXI) was constructed and put into operation in the WAK plant. As the operational performance of these two new items of equipment was very satisfactory - more than 50 t Pu have been put through up to now /36/ - it was decided to replace also the mixer settler in the first cycle by an electrochemical device. 1B-EMMA was designed and constructed in 1983, tested and installed in the WAK plant in 1984 and went to operation early in 1985.

Development of electrochemical columns started in 1975. Early devices were made of glass and tested in the UTE rig. In carrying out this development, an electrochemical column (ELKE) of 8 m length and 90 mm diameter was constructed. The outer tubing and the sieve plates, both serving as the cathode, were made of titanium, while the anode, a central rod, was made of platinated tantalum. Extensive work has been done on the applicability of hafnium, which has a high capture cross section rate for neutrons. A ROXI cell, whose containment is manufactured from hafnium, is in operation in PUTE and has given very satisfactory results.

EMMA, ELKE and ROXI cells on both the pilot plant and industrial scale have proved their operationanl power and reliability /37,8/. Titanium and hafnium have given satisfactory results as cathode material in respect of corrosion properties under the conditions of electrochemical REDOX reactions. The corrosion rates are smaller than 10 µm/y for both metals. For the anode, platinum has proved to be suitable (corrosion rate < 5 µm/y), and, for cost reduction purposes, is plated on the carrier tantalum.

In the seventies, a considerable amount of effort has gone into the development of centrifugal extractors, and a 6 stage version has been constructed on the pilot plant scale /38/. At the present time, centrifugal contactors are under construction to be built into the MILLI as an alternative 1st-cycle extractor.

OFF-GAS TREATMENT

Apart from iodine-129, the content of radioactive gases (Kr, C-14, T) in the off-gas of reprocessing plants is low. The successful retention of I-129 has been demonstrated on the technical scale and is considered as a reliable technology.

Nevertheless, processes have been developed up to the pilot plant scale for the retention of Kr. Two processes to separate Kr were developed: (1) the cryogenic rectification /39/ and (2) the R-12 wash at pressures slightly below atmospheric /40/. Work on the cryogenic rectification was started in 1976. A pilot plant (50 m^3/h)-KRETA was set up, operating with simulated off-gas. Prior to cryogenic rectification, traces of NO_2 and O_2 in the off-gas must be reduced with hydrogen in a catalytic process - REDUCTION /41/, and H_2O and CO_2 are removed in the adsorption process -ADAMO. The cryogenic rectification is performed at 5 bar to avoid Xe precipitation. This process is comparatively complicated due to the thermodynamic properties of the three gases N_2, Kr and Xe, involved in the process. There are some rather critical points in this process: the high Kr inventory, ozone formation and accumulation. Also it seems that the equipment requires special care during plant operation.

For these reasons R&D work was started in 1980 on the alternative freon absorption process, for the separation of krypton /40/. In this process, which works at pressures slightly below atmospheric, NO_x is absorbed with cold nitric acid /42/while the other impurity gases are ineffective in the absorption process. Xe and Kr are consecutively absorbed by lowering the temperature one after the other in 3-stage columns. The pilot plant demonstration of this process is scheduled within the next year. The absorption process is considered to be more flexible than the cryogenic process and the Kr inventory is at least a hundred times smaller.

REMOTE MAINTENANCE

Development work on remote maintenance and replacement techniques is being carried out at KfK's Engineering Department. It is expected that a reprocessing plant maintenance concept based on remote replacement of failed equipment /46,47/ will result in an improved plant availability due to a decrease of decontamination and repair times needed. Additional investments for remote handling equipment and installations should be counterbalanced by a decreased need of equipment redundancy, and by a more compact hot-cell layout since the process equipment can be installed in standardized racks alongside the walls of a large single process cell, instead of a multitude of small single-purpose cells. Connections and disconnections of piping, electrical leads, etc. are performed with the aid of a TV-controlled bridge crane system equipped with remote tools. Remote operation of certain equipment inside the cell, as for instance sampling stations, can be done by conventional masterslave manipulators and direct viewing through shielding windown. This technology will be applied and tested in the PAMELA high-level waste vitrification plant at Mol, Belgium, which is presently undergoing cold trial operation /48/.

OUTLOOK

Though the chemistry of the Purex process is rather simple, its reproduction in a technical plant is comparatively complicated. Usually the flowsheet consists of 5 extraction cycles, each containing 3-4 extractors and in addition a number of rework equipment items. As any disposal from the plant is severely restricted, organic and aqueous streams must be recycled. Chiefly, two requirements complicate the technology: (1) the high recovery yield for plutonium and (2) the high purity required for both products plutonium and uranium.

The future prospect for breeder technology relies to a great extent on its economics. The fuel recycle plays a primary role in this respect. Very important also is the interaction of reprocessing and fuel refabrication, characterized mainly by product specifications. When remote fuel refabrication is available and the gamma activity specifications decided, major simplifications will be possible. It seems, however, that remote refabrication will not come in the near future, and of course, some of the money saved in reprocessing must then be used for remote and shielding equipment for the fabrication process. Hence it seems necessary to discuss these questions in much more detail.

There is agreement that the uranium content in the plutonium product can be as high as 50% of the heavy metal without affecting the present refabrication techniques, and this also is of some help in compressing the flow sheet. If, in this connection, optimal condition in the extraction cycles can be maintained, electrochemical processes for plutonium reduction and oxidation are employed, and the extraction equipment is constructed requiring only short contact times in the high active part, model simulations very clearly show that both final purification cycles can be omitted.

For further simplifying the Purex process, one aim is to replace the purification cycles by other unit operations, e.g. crystallization processes, which are considered to be simpler in equipment and operation, but as effective as extraction. For some time to come, the Purex partitioning cycle will be retained in nuclear fuel reprocessing; considerable progress, however, can be achieved by selecting new and more economic unit operations prior to and after the high active extraction cycle.

REFERENCES

/1/ W.Marth, in F. Baumgärtner, K. Ebert, E. Gelfort and K. H. Lieser, Eds., Nuclear Fuel Cycle, Vol. 2, p. 5. Verlag Chemie, Weinheim (1983).

/2/ K. Ebert, in Report KfK 3740 (1984), p. 91.

/3/ F. Baumgärtner and W. Ochsenfeld, Report KfK 2301 (1976).

/4/ W. Ochsenfeld, F. Baumgärtner, U. Bauder, H.-J. Bleyl, D. Ertel and G. Koch, Proc. ISEC'77, p. 607. CIM Spec. Vol 21, Canadian Inst. Mining Metallurgy, Montreal (1979).

/5/ W. Ochsenfeld and H.-J. Bleyl, Atomenergie-Kerntech. 33, 251 (1979).

/6/ W. Ochsenfeld, H.-J. Bleyl, D. Ertel, F. Heil and G. Petrich, Proc. Symp. Fast Reactor Fuel Reprocessing, p. 77. Soc. Chem. Ind., London (1980).

/7/ W. Ochsenfeld, Proc. Symp. Fast Reactor Fuel Cycle, p. 349, British Nucl. Energy Soc., London (1981).

/8/ H. Schmieder, H. Goldacker, M. Heilgeist and G. Petrich, Proc. Symp. Fast Reactor Fuel Cycle, p. 281. British Nucl. Energy Soc., London (1981).

/9/ W. Ochsenfeld, H. Schmieder and S. Theiss, Report KfK 911 (1970).

/10/ F. Baumgärtner, E. Schwind and P. Schlosser, German Pat. No. 1905519 (1970).

/11/ H. Schmieder, F. Baumgärtner, H. Goldacker and H. Hausberger, Proc. Reaktortagung, p. 706. Deutsches Atomforum, Bonn (1971).

/12/ H. Schmieder, F. Baumgärtner, H. Goldacker, H. Hausberger and E. Warnecke, Proc. ISEC'74, p. 1997. Soc. Chem. Ind., London (1974).

/13/ H. Schmieder and K.L. Huppert, in F. Baumgärtner, Ed., Chemie der Nuklearen Entsorgung, Vol. 2, p. 50. Verlag K. Thiemig, München (1978).

/14/ H. Schmieder and L. Stieglitz, German Pat. 2449589-C2 (1974).

/15/ M. Heilgeist, H. Goldacker, H. Schmieder, G. Kreysa, G. Breitenbach and E. Heitz, Report KfK 2940 (1980).

/16/ W. Wagner and W. Ochsenfeld, in F. Baumgärtner, K. Ebert, E. Gelfort and K.H. Lieser, Eds., Nuclear Fuel Cycle, Vol. 1, p. 109. Verlag Chemie, Weinheim (1981).

/17/ W. Ochsenfeld, G. Petrich, H.-J. Schmidt, A.H. Stollenwerk and H. Wiese, Sep. Sci. Technol. 18, 685 (1983).

/18/ B. Kanellakopulos and G.P. König, Radiochim. Acta 33, 169 (1983).

/19/ N. Boukis, Thesis, Univ. of Heidelberg, Germany (1985).

/20/ W. Ochsenfeld, G. Baumgärtel and H. Schmieder, in A.S. Kertes and Y. Marcus, Solvent Extraction Research, p. 373. Wiley Interscience, New York (1969).

/21/ G. Petrich and Z. Kolarik, Report KfK 3080 (1981).

/22/ F. Baumgärtner and L. Finsterwalder, J. Phys. Chem. 74, 108 (1970).

/23/ G. Petrich, U. Galla, H. Goldacker, M. Heilgeist, M. Kluth, R. Schlenker, H. Schmieder and K. Ebert, Proc. Symp. Extraction'84, Dounreay, UK, in press.
/24/ G. Petrich and H. Schmieder, Report KfK 3740 (1984).
/25/ H.-J. Bleyl, in Report KfK 3775 (1984), p. 123.
/26/ R. Becker, L. Stieglitz and H. Bautz, Report KfK 3836 (1984).
/27/ L. Stieglitz and R. Becker, in F. Baumgärtner, K. Ebert, E. Gelfort and K.H. Lieser, Eds., Nuclear Fuel Cycle, Vol. 2, p. 333. Verlag Chemie, Weinheim (1983).
/28/ L. Stieglitz, W. Ochsenfeld and H. Schmieder, Report KfK 691 (1968).
/29/ H. Eschrich and W. Ochsenfeld, Sep. Sci. Technol. 15, 697 (1980).
/30/ W. Ochsenfeld and J. Schön, Proc. ISEC'83, Denver (1983), p. 92.
/31/ D. Ertel, in F. Baumgärtner, K. Ebert, E. Gelfort and K.H. Lieser, Eds., Nuclear Fuel Cycle, Vol. 2, p. 229. Verlag Chemie, Weinheim (1983).
/32/ P. Groll and J. Römer, in F. Baumgärtner, K. Ebert, E. Gelfort and K.H. Lieser, Eds., Nuclear Fuel Cycle, Vol. 2, p. 257. Verlag Chemie, Weinheim (1983).
/33/ P. Groll, Proc. 14th Europ. Symp. Computer Control Oper. Chem. Plants, Laxenburg. EFCE Publ. Ser. No. 17 (1981).
/34/ P. Groll, M. Persohn, L. Röder, J. Römer and P. Schuler, Report KfK 3533 (1983).
/35/ H. Goldacker, H. Hausberger, H. Schmieder and H. Wiese, Proc. ANS Conf. Fuel Cycles for the 80's, Gatlinburg, USA, CONF-800943-(Summ.) (1980), p. 104.
/36/ H. Goldacker, H. Schmieder, U. Galla, M. Kluth, R. Schlenker and H. Evers, Proc. ISEC'83, Denver (1983), p. 329.
/37/ H. Schmieder, H. Goldacker, U. Galla, M. Heilgeist, M. Kluth and G. Petrich, presented at ANS Int. Meeting Fuel Reproc. Waste Management, Jackson Hole, USA (1984).
/38/ B.F. Roth, Report KfK 862 (1964).
/39/ R. von Ammon, H.-G. Burkhardt, E. Hutter and G. Neffe, Ber. Bunsenges. Phys. Chem. 83, 1143 (1979).
/40/ E. Henrich, R. Hüfner, F. Weirich, W. Bumiller and A. Wolff, Proc. 18th DOE Nucl. Airborne Waste Management Air Cleaning Conf., Baltimore, USA (1984), Paper No. 14/3.
/41/ R. von Ammon, G. Knittel and E. Hutter, in Management of Gaseous Wastes from Nuclear Facilities, IAEA-SM-245/12 (1980), p. 229.
/42/ E. Henrich, U. Bauder, H.-J. Steinhardt and W. Bumiller, Proc. 18th DOE Nucl. Airborne Waste Management Air Cleaning Conf., Baltimore, USA (1984), Paper No. 6/6.
/43/ E. Henrich, T. Fritsch, A. Wolff and H.-J. Schmidt, KfK-Nachr. 14, 109 (1982).
/44/ R.D. Penzhorn, H.E. Noppel, P. Schuster, H. Leitzig, A. Dorea and R. Kröbel, Proc. ANS Int. Meeting Fuel Reproc. Waste Management, Jackson Hole, USA (1984).

/45/ H. Enderlein and W. Kathol, in Report KfK 3278 (1981) p. 56.
/46/ G. Böhme and H. Frotscher, in Report KfK 3775 (1984) p. 171.
/47/ G. Böhme et al., presented at IAEA Seminar Remote Handling Equip. Nucl. Fuel Cycle Facil., Harwell, UK, Oct. 2-5 (1984).
/48/ M. Selig, Proc. 31st Conf. Remote System Technol, Detroit, USA, Vol. 1 p. 131 (1983).

The Current Status of Fast Reactor Fuel Reprocessing Programme in India and the Future Plans

G.R. Balasubramanian, Reprocessing Development Laboratory, Reactor Research Centre, Kalpakkam, INDIA

The Indian nuclear programme aims at making a significant contribution to the electric power generated in the country and envisages an installed capacity of 1000 MW by the end of the century, flowing from a series of pressurised heavy water type of reactors. It lays emphasis on utilisation of the fast breeder power reactor as a second phase, availing the plutonium generated. A comprehensive scheme for developing the fast reactor and its fuel cycle has been launched by creation of a dedicated research centre at Kalpakkam which coinsists of a fast breeder test reactor and various research laboratories associated with the fuel cycle. The test reactor is likely to be commissioned by the middle of this year and other laboratories are already operational. Work on design of a prototype 500 MW fast reactor is also in an advanced stage. The Centre has identified development of reprocessing technology as one of the important objectives and has evolved an organised approach. As a first step a reprocessing development laboratory has been commissioned and operation for the past few years. The scope and details of this facility has been described in the paper (1) presented in 1981 at London. This paper describes the salient aspects of the activities carried out so far in various areas and also its future planning for setting up of reprocessing facilities for reprocessing of fuels from its future proposed prototype fast reactor.

The reprocessing laboratory has almost completed its task of the development of equipment needed for reprocessing fuel from the test reactor. The philosophy adopted for the design of the equipment, follows the general guidelines listed below:

THE EQUIPMENT

(1) must have reliable performance with maximum efficiency for the process function
(2) must be made of material having good mechanical strength and corrosion resistance to ensure long life
(3) must be fabricated to the highest quality standards of nuclear chemical plants
(4) must be suitable for prolonged remote operation without maintenance or where maintenance is inescapable must be amenable for remote maintenance
(5) must have low hold-up volume to minimise inventory of fissile material
(6) should be designed to be critically safe for all operating conditions
(7) must have ease of decontamination and replacement
(8) must be easy for scaling-up
(9) must be of low cost.

The equipment is designed and a prototype is fabricated in the first instance. Subsequently, it is tested with the uranium-nitric acid system under simulated conditions in the engineering laboratory which forms part of the laboratory. After extensive evaluation during this prolonged operation for their individual performance, for parameters such as ease of operation and control, they are integrated into operations consisting of head-end and one cycle of solvent extraction. During this prolonged operation under simulated conditions in the cells, the equipment is evaluated for its performance with respect to its reliability, guaranteed performance of process efficiency, trouble-free operation, ease of remote maintenance, matching performance with the related equipment. Once the equipment has been validated in this rig, the equipment design is accepted for use in a hot pilot facility where the fuel from the experimental fast reactor is to be treated. Experimental reprocessing of the test reactor fuel will provide the necessary information and experience for designing a scaled-up version of the equipment for the reprocessing for the prototype power fast reactor programme in the future.

Development work carried out on some of the process equipment is highlighted in the following sections.

Single pin chopper

As the process design adopts a chop-leach process at the head-end step and the fuel is going to be dismantled at the reactor site, it was decided to have a single pin chopper in the first instance. The chopper will be remotely operated. The fuel pins from a magazine, which has a capacity for 10 pins, are fed one by one to the chopper. The chopper has a pneumatically operated cutting blade system and gripping device. Each pin is fed into the chopper by means of a stepper motor. The blades of the chopper and gripping blocks are so designed as to be removed remotely by Model-8 manipulators. Individual components are designed such that their weight is well within the handling capacity of the manipulators. The chopper is operated by a microprocessor based programme controller. After the completion of one pin, the push rod retracks and the magazine is suitably rotated to align the position of the next fuel pin in it with the push rod. The size of the chopped pieces can be regulated by controlling the feed rate. The force on the pneumatic cylinders is also variable by adjusting the air pressure. The delivery system to the dissolver is suitably designed such that the chopped pin can fall smoothly into the dissolver. There is also a separate extraction system to collect any dust produced.

Dissolvers

The thermosyphon type dissolver of eversafe geometry and 2kg batch capacity was developed. The basket and hull discharge systems are designed in such a way to minimise contamination in the cell, during handling. The recirculation in the dissolver can be established with nitrogen pumping.

It is proposed to use a mixed carbide fuel in the experimental reactor. One of the methods proposed for dissolution is electrolytic oxidative dissolution. The dissolution of fuel from this reactor will be carried out in an electrolytic dissolver. The electrolytic oxidation in addition to accelerating the kinetics of Pu dissolution, oxidises the organic acids formed. A dissolver of 100g scale has been tested with uranium carbide-nitric acid system. Tests on corrosion rates of electrodes for the dissolver have also been carried out. An anode using titanium with substrate coating is found to function well. Titanium is considered as the material of construction for the dissolver.

Off-gas scrubbers using mercury nitrate and strong nitric acid are being used for treatment of the off-gases. It is proposed to extend the electrolytic process for the oxidation of the Pu for conditioning its valency state suitable for solvent extraction.

Feed clarification

In view of the presence of insoluble fission products in the dissolver solution a feed clarification step has become essential in reprocessing the fast reactor fuel. A solid-liquid separation system and one utilising a centrifuge are being developed in parallel. The filter element is of tubular geometry and made of sintered SS media with a pore size ranging from 2-5 μ. The element can be removed using a pneumatic cylinder. A pulsing system is incorporated to dechoke the filter during part of the cycle to prolong its life. Vacuum is used on both the cycles of filtration. The filter element can be removed along with the fission product insolubles. Insolubles can also be collected as slurry and can be passed on to a separate tank for subsequent disposal. Rates ranging from 10-80 litres per hour have been achieved using 3" diameter elements. In addition, the sintered filters have been tested for remote replacement of the cartridge. A centrifuge has been identified as the alternative system for feed clarification. A centrifuge of bowl length 1 metre and diameter 3" has been tested for its separation efficiency.

Solvent extraction contactors

A wide range of solvent extraction contactors have been developed for different applications. For use in the very highly active first cycle, it was decided to develop centrifugal contactors because of their low residence time. In the first instance, single stage centrifugal contactors were developed. The development started with a 5cm diameter bowl, 200ml hold-up single stage contactor. With this model hydraulic characteristics and basic aspects of design etc., were studied. Subsequently for immediate requirements, the need became apparent for a small throughput, low hold-up contactor. A tubular type of extractor was chosen for this purpose. With the combined hold-up of mixer and settler of around 15cc, the contactor would easily provide the required characteristics. The contactor gave a good mass transfer efficiency and good capacity as well as required characteristics of low hold-up. This development also paved the way for the next stage of development, a multi-stage contactor. The experience with the single stage contactor also enabled modification of the single stage design into an annular type design. In the next stage of development attempts are being made to replace the electrical drive system with a pneumatic system. Utilisation of this contactor in a hot cell and the possibility of maintaining it were all checked in a mock-up facility. Containment systems using hydraulic seals were also carried out. The annular stage contactors are at present designed to be tried as a bank of contactors using 8 stages. Problems encountered in getting suitable seals and establishing hydraulic characteristics have been overcome.

Air-pulsed mixer-settlers

Air-pulsed mixer-settlers are also being developed for processing the thorium blanket. These mixer-settlers are of eversafe geometry with the mixing achieved by two oscillating air columns. The interstage liquid movement is achieved by air lifts and the interface in individual stages can be controlled by remotely operable weirs. A detailed study was carried out for the design of the probes and different probe geometries were tried. The pulsing system also produced a system of recirculation. An empirical correlation for sizing the probe was also arrived at.

Horizontal baffle contactor

A horizontal baffle contactor which was initially designed for solvent wash applications indicated good characteristics suitable even for the first cycle of extraction. The contactor has the advantage of eliminating complicated aqueous ports. It has good efficiency and the throughput is much better than the mixer-settler. This equipment is being studied extensively and is being considered for various applications. In this case also, mixing is by means of oscillating air columns.

Already enough experience is gained in operating pulsed columns for thermal reactor fuel reprocessing, and it is anticipated that at an appropriate time, use of pulsed columns could also be extended for this application. Studies on using the mixer-settlers with in-situ electrolytic reduction for partitioning of plutonium from uranium have commenced. Studies carried out on a multi-limb, pulsed vertical column show promising results with respect to its efficiency and its performance is under detailed investigation.

AUXILIARY SYSTEMS

Metering systems

Because of the problems of maintenance posed by the diaphragm type metering pumps, work on metering systems requiring little maintenance with precision comparable to that of mechanical pumps was pursued. Two systems have been studied. First, a constant level variable orifice device. In this a constant level is maintained in a tube and the liquid from this is withdrawn through a fixed orifice. With a good level control system, the height of column of liquid above the orifice was maintained constant at a chosen level. For a fixed level and orifice the flow is constant within limits. Second, a device with constant liquid level from where liquid is made to flow through a variable orifice was also tested. This device with suitable choice of height and orifice can cover a wide range of liquid flows. Since these are amenable for control, they are good devices for automatic control of the solvent contactors as well.

Systems for taking samples and pneumatically conveying them to the laboratory for analysis were also studied. From this work, pneumatic sample despatch and receipt stations were developed.

Other equipment systems developed include alpha, beta and gamma casks for the active material transport between the cells, pneumatically propelled intercell trollies etc.

In order to ensure that each piece of equipment developed meets the overall demands with respect to performance and reliability for a reprocessing plant, they are rigged into a circuit consisting of head-end and one cycle solvent extraction operation, using the U-TBP-nitric acid system. The units are tested for their process performance, reliability characteristics, remote operability, etc. The experience gained in this is very useful for modifications of designs for the hot pilot runs. In addition to the equipment testing, it is also proposed to test new plant design concepts. Plants now operating are mainly designed on the basis of remote operation and direct maintenance. The current thinking for the design of the fast reactor fuel reprocessing plant is to adopt a system which is a combination of remote operation, remote maintenance and direct maintenance. The equipment which demands least attention and has low probability of failure is located in cells which are designed for remote operation and direct maintenance. Other equipment such as solvent extraction contactors, sampling systems, etc. are separately located in a cell designed for remote operation and remote maintenance. The single cycle integrated operation of the equipment will also evaluate these concepts.

The laboratory studies carried out have helped to finalise the flow sheet for reprocessing mixed carbide fuel of high plutonium content. Direct dissolution and oxidation of organic acids using an electrolytically regenerated ceric system have been studied with unirradiated mixed carbide. Pyrohydrolysis has also been studied as an alternative system. Solvent extraction flowsheets have been formulated based on studies on extraction equilibrium, including those related to third phase formation.

Various control parameters have been evolved using a computer program. These have been verified by mixer-settler runs on laboratory scale.

A mini-plant to test the reprocessing of irradiated fuel from the test reactor using the schemes evolved is being got ready. It consists of a series of alpha contained lead shielded enclosures with processing capacity to meet the immediate requirements. The concrete cells which are functioning as an integrated test facility will be converted to a pilot plant for regular reprocessing of fuel, incorporating any necessary modifications indicated by the current single cycle operation.

Simultaneously, planning is in progress for developing the engineering for reprocessing fuel from the prototype reactor. As a part of a large reprocessing complex that is being constructed to reprocess fuel from the CANDU-type thermal reactors operating in the power stations located in the vicinity, a test plant to reprocess fuel from fast reactors is also planned. The plant will reprocess the fuel from the test reactor on a regular basis and carry out R&D functions for engineering development needed for designing larger plants. The cells are planned with facilities to have operations carried out on a progressively enhanced scale and will be a test bed for scaled-up versions of equipment.

With sufficient experience gained from these operations, the facility is also planned to reprocess on an experimental basis fuel discharged from the prototype reactor with construction of a suitable head-end facility. It is anticipated that by the end of this century, necessary expertise needed to reprocess fuel from power fast reactors will be generated. The present reprocessing development laboratory will be functioning to develop schemes to meet the reprocessing of Th fuels that are proposed to be used in heavy water type reactors as well as the blanket in the test fast reactor.

REFERENCES

1) G.R. Balasubramanian "Reprocessing of Fast Reactor Fuel and the Indian Approach to it". International Conference on Fast Reactor Fuel Cycle, London (1981).

2) P.G. Selvaraj, M. Venkataraman and G.R. Balasubramian on "Development of Equipment for Fast Reactor Fuel Reprocessing", Fuel Reprocessing and Waste Management Meeting, Jackson, Wyoming, USA (1984).

Experience and Trends in Irradiated Fuel Reprocessing in Italy

G. Grossi, A. Moccia, G. Rolandi, ENEA – Fuel Cycle Dept., Reprocessing Project, Italy

ITALIAN NUCLEAR POWER PROGRAM

The present situation in Italy, concerning the nuclear power production, is shown in Tab. 1, which gives the amount of spent fuel sofar accumulated and the expected yearly discharge for the reactors presently in operation, in construction and in the planning stage. One can see that only one new power station, is presently under construction: the Montalto power station, with two BWR units of 1000 MWe.

TABLE 1 Spent Fuel From Italian Power Stations

Reactor	Status	Power, MWe	Type	Commercial operation	Storage capacity, tU	Fuel discharge, tU/y	Discharged to 1984, tU
GARIGLIANO	shut down	160	BWR	1964	45	11	82
LATINA	in operation	160	GCR	1964	–	50	1000
TRINO	" "	270	PWR	1965	50	12	100
CAORSO	" "	894	BWR	1981	420	25	50
MONTALTO I and II	under construction	2 × 1000	BWR	1988-89	2 × 10 reloads	2 × 27	–
PIEMONTE	planned, site agreed	2 × 1000	unified PWR design	after 1995	2 × 10 reloads	2 × 22	–
LOMBARDIA	planned	2 × 1000	"	"	"	2 × 22	–
PUGLIA	planned	2 × 1000	"	"	"	2 × 22	–

Additional 6000 MWe of unified PWR design have been recently authorized by the National Energy Plan, approved by the Parliament on 1981.

NATIONAL POLICY CONCERNING REPROCESSING

2.1. The National Energy Plan, while authorizing these three new power stations, confirmed the **Italian** option towards the reprocessing of thermal spent fuel. Reprocessing is considered the best management concept for the spent fuel discharge from the domestic nuclear power stations, capable of "taking care" both of their fissile material content and of the fission products and transuranic isotopes produced. This target can be met provided that an advanced reprocessing techonology is adopted, in order to get improved conversion processing and handling procedure for final products (Pu or Pu and U), separate recovery of actinides from the bulk of fission products and, generally, optimisation of process schemes, effluents release and waste production.
On this basis, industrial reprocessing operations could be carried out, meeting, at the same time, the most severe requirements concerning radiological protection, safety and safeguards. The choice favorable to reprocessing is consistent on the other hand with the **Italian** positive option for fast breeder reactors.

2.2. The reference strategic frame envisaged is the reprocessing of the domestic spent thermal fuel in a domestic plant, but Italy's participation to some kind of multinational venture, on a European basis, as an alternative or as a complement to the domestic industrial plant, is also strongly envisaged.

2.3. There are, nevertheless, several conditions to be verified, before to decide any industrial undertakings on reprocessing. First of all the realisation of an industrial reprocessing plant in Italy will only be decided when the number of nuclear power stations in operation and construction will be at least 10 to 12, in order to justify a 400 tU/y plant, this being the minimum capacity considered economically viable. This condition sets the realisation of the plant, if any, after the year 2000.

But, in our view, it is the decision on the use of recovered

plutonium which plays the leading role: the recovered Pu is a "perishable" material: that is to say, its nuclear characteristics are "getting worse" rather rapidly with time; moreover, its handling and storage set up severe safety and safeguards problems.
It seems therefore preferable to store the spent fuel and to reprocess it only when the need forPu calls for it.
In connection to this, one has also to consider that reprocessing of spent fuel after a relatively long cooling time (say, 10-20 years), offers considerable advantages from the technical and radiological protection point of view. Finally, the economics of reprocessing must be deeply examined, in any case, before deciding the high capital investment required. It is well known that the very high cost of reprocessing entails a cost of recovered fissile materials much higher than the presently envisaged commercial value.

2.4. In conclusion, in the present Italian situation "delayed reprocessing" seems to be the best approach. As a consequence, a centralized Away From Reactor (AFR) storage facility has to be provided and a substantial R and D programme on reprocessing technology, has to be set up, together with consistent initiatives of industrial promotion in this field. All these activities will be briefly reviewed in the following.

ROLE OF NATIONAL OPERATORS

3.1. ENEL (Ente Nazionale Energia Elettrica) the National Electrical Energy Board, is the operator of the nuclear power stations, and, as such, it has to take care of the spent fuel management. It has fulfilled this task, up to now, in a flexible manner; in some cases the spent fuel has been reprocessed by contracts with BNFL in UK and COGEMA in France, in some others it has been stored.

3.2. ENI (Ente Nazionale Idrocarburi) the National Hydrocarbon Company, has been designated by the Gonvernment as the industrial operator for the fuel cycle activities. As such, it has undertaken several initiatives, in connection with the ENEL activities above described, through its subsidiary companies AGIP Nucleare (AGN) (now incorporated into AGIP) and NUOVO PIGNONE. As an example, transport cask (the AGN-1 cask) for the transfer of spent fuel from TRINO and GARIGLIANO, was designed, manufactured, qualified (with the help of ENEA), and delivered to ENEL.

TRENDS OF THE R AND D ACTIVITIES CARRIED OUT BY ENEA

4.1. A comparatively large R and D effort is being carried out and is foreseen in the future by ENEA, on the basis of the work done in this field in the past 25 years, starting with the participation at the Eurochemic venture, early in the 1960s.
The purpose of the ENEA R and D program is to implement and complete the know-how existing in Italy on reprocessing technology by an effective and coordinated use of the Eurex Pilot Plant at the Saluggia Center, the Hot Test Facility ITREC together with the Technological Hall at the Trisaia Center, the Research Laboratories at the Casaccia Center, near Rome, some University Institutes and the Industry.
A well finalized R and D program in the field is deemed not only consistent with the general option towards reprocessing, but also necessary in order to be ready either to construct the industrial plant on a national basis or to participate, in a qualified manner, to an international venture on reprocessing of thermal spent fuel.
We believe that an industrial reprocessing plant to be erected after the year 2000, will need important technological improvements, concerning not only the process but also the remote operation and maintenance concepts, process control and automation, fissile material accountability and safeguards.
Here too, we intend to carry out this program within the frame of collaboration agreements with the european countries advanced in this field.

4.2. The main facilities for this R and D program are as follows:

- The Eurex plant (Saluggia), a pilot plant of about 50 kgU/d capacity, which entered into hot operation in 1970 and has performed campaigns with MTR fuel (500 elements) and after necessary modifications and improvements (such as a full bundle shearing machine) has processed in 1981/83 1.7 tU of CANDU type oxide fuel, in the frame of a joint experiment ENEA-AECL.
- The ITREC Hot Testing Station of the Trisaia Center, which has been erected, originally, as an integrated reprocessing and remote refabrication unit for the Th-U cycle. The reprocessing section only has been put in operation in 1975, and a few Th-U reprocessing campaigns, have been carried out. An experience

complementary to that of EUREX was obtained, in particular as regards the remote maintenance and the modular rack conception.

ENEA UNDERTAKINGS ON INDUSTRIAL PROMOTION

ENEA, according to its institutional tasks, has to carry out, besides the R and D programs, and in connection with them, several industrial promotion undertakings.

This approach will also apply to the reprocessing field, where ENEA has to play a leading role in preparing the future industrial activities. The natural counterpart is, of course, the ENI group, which has already carried out a few significant realisation, already mentioned: such as, the AGN-1 cask.

Indeed, the most important near-term industrial undertaking concerning the "back-end" of the fuel cycle, will be the erection of an Away From Reactor Central Storage Facility, which will be needed for the middle of the '90s. This facility, will provide not only the storage capacity needed for the Caorso spent fuel, but also an additional capacity which will ensure a high operational flexibility for the Montalto and the other planned power stations. This central AFR storage installation has also a strategic meaning, insofar as it actually permits "delayed reprocessing". In this respect, it could be considered and designed as connected to the possible future reprocessing plant.

ENEA and ENI have recently decided to set up a joint process engineering company, whose task will be to concentrate all the design capabilities of the two organisations, in order to collect, review, qualify and organize all the reprocessing technology know-how so far gained.

The actual work to be carried out by this new company will be as follows:

- the conceptual and detailed design of the AFR storage facility and the related fuel transport and handling equipment;
- the conceptual design of the industrial reprocessing plant, mainly in order to provide a reference guide for the definition and finalisation of the R and D work described above;
- assistance to the ENEA's experimental reprocessing facilities, for any new modifications and improvements.

This joint company should constitute a flexible instrument capable of preparing any possible future reprocessing industrial activities in Italy.

PRESENT SITUATION AND TENDENCIES OF THE REPROCESSING R AND D IN ITALY

6.1. The big R and D effort in the reprocessing technique has already obtained in the past some results of significant importance.
The main tasks to be mentioned in this field are:

- the development of the tertiary amine based flowsheet for the reprocessing of MTR-type irradiated fuels at EUREX plant (selective extraction of enriched uranium);
- the development of the flowsheet for the reprocessing of CANDU irradiated fuels at EUREX plant (selective extraction of plutonium);
- the development of a flowsheet for the separation of uranium from thorium using branched-chain amides as new extracting agents, to be applied in the ITREC plant;
- the development of complexing compounds, hydroxamic acids, useful for final product purification;
- the development of TDR technique for liquid level measurement;
- the development of densito-conductivity technique for on line control of acidity and total actinides concentration;
- the development of X-ray absorptiometry techniques for SNM concentration measurement in process control;
- the development of on line alpha monitoring system for plutoniom losses detection in liquid streams;
- the development of gamma spectrometry techniques for plutonium determination.

The present R and D activities are now oriented on the following two main lines, according to the ENEA Five Years Plans Guidelines:

1) to get acquainted, and further to optimize, the already well established processes and technologies;
2) to develop more advanced processes and technologies as possible alternatives to the methods currently used.

The main fields in which these R and D activities are developped are:
- separation chemistry of actinides and fission products;
- advanced NDA-based off-line and on-line instrumentation technique;

- plant demonstrations of processes, components and instrumentation;
- treatment and conditioning of radioactive wastes (not discussed in the present paper).

6.2. As well known, the PUREX process is nowadays the most proved and accepted process for the reprocessing of irradiated fuels, both in pilot and in industrial scale.
However, there are many steps of this process that need a further improvement, in order both to minimize the production of radioactive wastes, and to overcome the problems originated by the very high burnup fuels coming from the last generation nuclear power plants.
It is important, in this connection, to point out the problems originated by the adaptation of PUREX process to the reprocessing of fast reactors fuels.
Some of the most significant R and D items on this subject are:
- the development of advanced chemical methods for plutonium-uranium partition and plutonium stripping (organic-soluble hydroquinones; hydroxylamine nitrate),
- studies on the chemical and radiolytical stability of diluent and extractants, and on solvent regeneration procedures,
- development of advanced components (centrifugal extractors - to minimize radiolysis effect; electrolytic pulsed column - to minimize waste production),
- the development of advanced flowsheets for the "pushed reprocessing" (separation of all actinides: U, Np, Pu, Am, Cm from fission products).

6.3. Starting from reasonable concepts related to the fast breeder fuel cycle, a number of advanced R and D lines is at present pursued at ENEA labs and facilities.
Main tasks are represented by:

i) Flow sheet studies with a final product constituted by U-Pu nitrate mixed solution.
ii) Selection of high-performance amidic extractants.
iii) Automatic control for obtaining reproducible batches of final products with given Pu enrichment.
iv) Plant verification of real or simulated process schemes.
v) General problems of extraction equipment automation.
vi) Development of remotized coprecipitation process via sol-gel technique.
vii) Evaluation of special separation techniques.

- Task i): The coprocessing flow-sheets are selected in the range of unstable operation. Both options for double or single extractant process are followed. The key problem is to obtain high Pu enrichment factors together with low Pu losses.

- Task ii): In the field of improved extraction systems, highly U - selective extractants are under study, and could be available.

The most important problem is to improve the loading capacity and avoid third-phase formation.

- Task iii): In connection with the selection of unstable operation (see task i), an automatic control of the operation is mandatory.

- Task iv): Hot test in EUREX plant (next Trino campaign) and cold test in the technological hall of the Trisaia Center (tail-end treatment for simulated Pu enrichment), are foreseen. The characteristics of the two plants highly facilitate the verification of coprocessing schemes.

- Task v): An algorithmic approach for evaluation of different automation modes to be applied to the available solvent extraction codes, is foreseen.

- Task vi): The integrated reprocessing - coconversion process applied to U-Th fuels is the first goal envisaged for the hot demonstration of FBR cycle in ITREC plant.

- Task vii): Ion exhange, reversed phase chromatography and Supported Liquid Membranes are under examination.

Most of above mentioned research lines are of interest for the LWR fuel cycle and for a delayed planning of a national or joint reprocessing plant. For istance, the possible suppression of the use of any type of reductant in U - Pu partial or total partition is thought to determine a renewed interest in a broader use of centrifugal contactors for large-capacity plants. Also the overall plant design could be highly simplified by the adoption of coprocessing schemes.

Status of the Breeder Fuel Cycle in the United States*

William D. Burch, Director, Consolidated Fuel Reprocessing Program, Oak Ridge National Laboratory, Oak Ridge, Tennessee 37831

INTRODUCTION

This paper reviews the status and plans for the fast reactor fuel cycle in the United States. The United States is undertaking a complete reexamination of its entire breeder program strategy, and the direction of the new program is not yet clear. In this paper, I will describe studies in progress to examine the associated fuel cycle strategies as they relate to the overall emerging breeder strategy. The present status of and recent developments in the fuel cycle R&D programs will also be summarized.

STATUS OF REPROCESSING PROGRAM

Over the past decade the United States has carried out an extensive program to develop advanced technology for breeder reprocessing while examining a range of project options to close the fuel cycle. During the mid-1970s, when a large number of breeders were envisioned on-line by the end of this century, a Hot Experimental Facility was conceptually designed to recycle fuel from the

---

*This research was sponsored by the Office of Spent Fuel Management and Reprocessing Systems, U.S. Department of Energy, under Contract No. DE-AC05-84OR21400 with Martin Marietta Energy Systems, Inc.

first four to six breeders. With the slowdown of the programs, options for closing the fuel cycle on a smaller scale (consistent with the perceived needs of the planned reactors) were examined. About three years ago, we began examining a project called the Breeder Reprocessing Engineering Test, or BRET. This project would utilize the major hot-cell space in the Fuels and Materials Examination Facility (FMEF), which also houses the Secure Automated Fabrication (SAF) line. The FMEF, now completed at the Hanford Engineering Development Laboratory (HEDL), is operated by the Westinghouse Hanford Company. In the originally planned first phase of the BRET project, fuel from the Fast Flux Test Facility (FFTF) would be recycled. With overall funding constraints on civilian nuclear power development in the United States, BRET has been deferred and is being reexamined in the context of the overall emerging U.S. breeder strategy. A primary objective of BRET, final development and testing of the concepts under development in a real, "hot" environment, could still be accomplished using only FFTF fuel. Decisions to continue with BRET to accomplish this role must await the completion of the entire breeder strategy review process.

ELEMENTS OF FUEL CYCLE STRATEGIC PLANNING

Options for closing the breeder fuel cycle are being reexamined bringing together the planning activities of all organizations directly or indirectly involved: (1) The U.S. Department of Energy's (DOE's) Consolidated Fuel Reprocessing Program, which I represent; (2) the fabrication program interests at HEDL; (3) the General Electric Company and Rockwell International, the two reactor vendors chosen by the DOE to carry out the innovative concept studies over a three-year period; and (4) the breeder Consolidated Management Office (CoMO).

Potential Reactor Scenarios

In order to define the fuel cycle study, it was necessary to project reactor scenarios which would help define fuel recycle requirements. As the reactor strategies evolve, the fuel cycle plans will be made compatible with the reactor projects. For this purpose only, the following were initially assumed:

1. The FFTF will continue operating.
2. A small, prototype advanced liquid metal reactor comes on-line at Hanford in 1995.
3. One or more reactor vendors provide a commercial breeder at a utility site by the year 2000.

It must be emphasized that these assumptions in no way reflect any judgments or decisions as to the direction of the reactor strategies.

The purpose of this strategy planning thus becomes to develop options for dealing with one or more of the above-assumed reactor scenarios, recognizing that all of them may not come about but providing a sufficiently flexible approach to accommodate those that do.

Fuel Cycle Options

Three potential fuel cycle options that are focused only on near-term needs have been defined to date:

1. providing a co-located fuel cycle facility for reactor scenario 3;
2. utilization of BRET/SAF in an expanded mode as needed; and
3. delayed (five to ten years) closure, initially relying on presumed available supplies of plutonium, until the fuel cycle could be closed later for more than one reactor.

For the second option we have examined the expansion capability of BRET and the requirements for transportion of spent breeder fuel for an off-site reactor. Although the study is only partially completed at the time of the written paper, at least some of the results should be available at the time of the symposium. Early conceptual work shows that BRET could likely be expanded to accommodate up to 2000-MW(e) breeder capacity.

Working with the reactor vendors, scoping concepts are being developed for small, co-located facilities to close the fuel cycle at a utility reactor site. The primary benefits from such an option stem from the simplicity of managing the entire reactor–fuel cycle complex within a single organization and the perception that eliminating all transportation of plutonium will reduce costs and minimize safeguards concerns. During this year we are exploring this proposal in depth, developing a sounder basis for costs, determining the sensitivity of capacity of costs, and the overall feasibility. The reactor vendors are utilizing the data provided in these fuel cycle studies to develop an on-site fuel cycle strategy for their entire innovative study. These studies will compare the co-located option with the BRET expansion mode. We expect to examine a limited range of capacity for the co-located facilities since it might be necessary to address stations with outputs from single 400-MW(e) reactors to multiple units that might eventually total 2000 to 3000 MW(e). Since a great deal of engineering and cost information has been developed from the work done on BRET and SAF and since the co-located facilities are in the same range of size, a reasonable picture of costs for a range of capacities should be obtained..

Plans developed within CoMO for an initial reactor project have envisioned that cost savings could be realized by delaying closure of the fuel cycle as long as supplies of plutonium could be obtained relatively inexpensively. This might prove to be only five to ten years, but even that period might be long enough for the costs to be spread over more than one reactor rather than loaded on the initial project. This concept will be explored further, although it will undoubtedly prove quite difficult to estimate how much plutonium will be available and at what price. This study is also briefly addressing the future coupling of an LWR reprocessing industry for plutonium supply to breeder recycle.

In evaluating and comparing these options, efforts will be focused primarily in two areas: (1) costs and (2) the practicality of closing the fuel cycle by such an option.

STATUS OF DEVELOPMENT PROGRAMS

Some six years ago a project known as the Integrated Equipment Test (IET) facility was started at the Oak Ridge National Laboratory (ORNL) to provide a test bed for the development of breeder reprocessing technology. This facility has been upgraded significantly as new concepts and hardware have been devised, and it is now complete and operational. Figure 1 shows the Remote

FIGURE 1 Remote operations and maintenance development of the IET. From Oak Ridge National Laboratory

Operations and Maintenance Development (ROMD) area where the major head-end equipment of a breeder reprocessing plant is installed in a simulated process cell, complete with overhead handling systems for remote maintenance. Included is the CRL Model M-2 servomanipulator as well as two cranes and two power manipulators. Also visible are the laser-based disassembly-feed system, a whole-bundle shear, and the ORNL-developed rotary continuous dissolver. In another area a single cycle of solvent extraction, using centrifugal contactors for extraction and scrub, is used to develop the contactors and to recover the uranium from the dissolver product. Remote maintenance tests have been performed on much of this equipment, and such tests will continue.

Over the past few years the breeder reprocessing program was refocused on smaller-scale equipment as needed for BRET. The remote maintenance concepts are largely independent of the scale of the process equipment, and the development activities remain largely unchanged. On the other hand, the process equipment had to be scaled down from the earlier work. Two prototype banks of smaller centrifugal contactors have now been operated successfully, and we are embarking on a longer-term reliability test of those prototypes.

To depict more clearly the concepts for BRET, a cross section of the entire facility is given in Fig. 2, and a plan and cross section of the Main Process Cell (MPC) are shown in Figs. 3 and 4. The cell plan view (Fig. 3) shows that

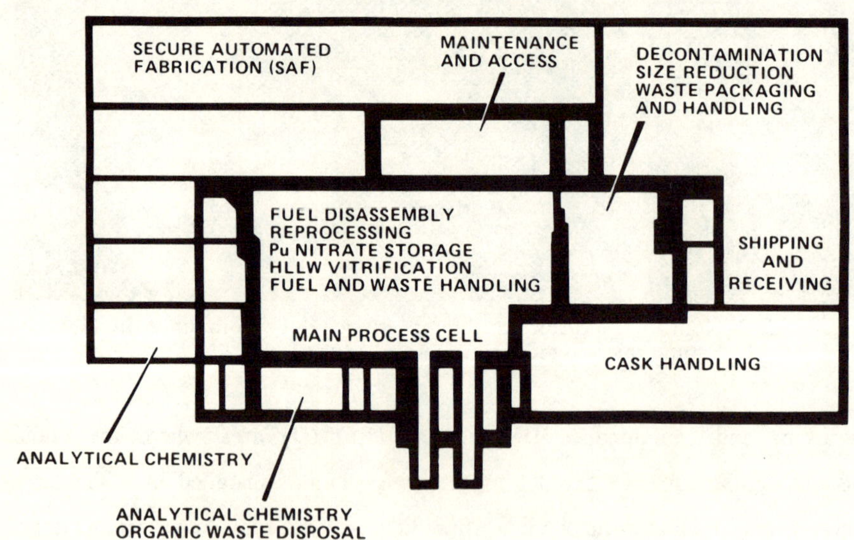

FIGURE 2 Cross section of the proposed BRET facility.
From Oak Ridge National Laboratory

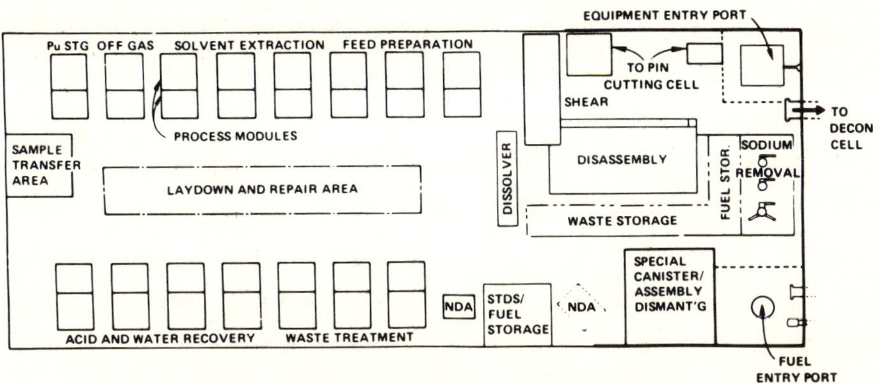

FIGURE 3 Plan of main process cell of BRET.
From Oak Ridge National Laboratory

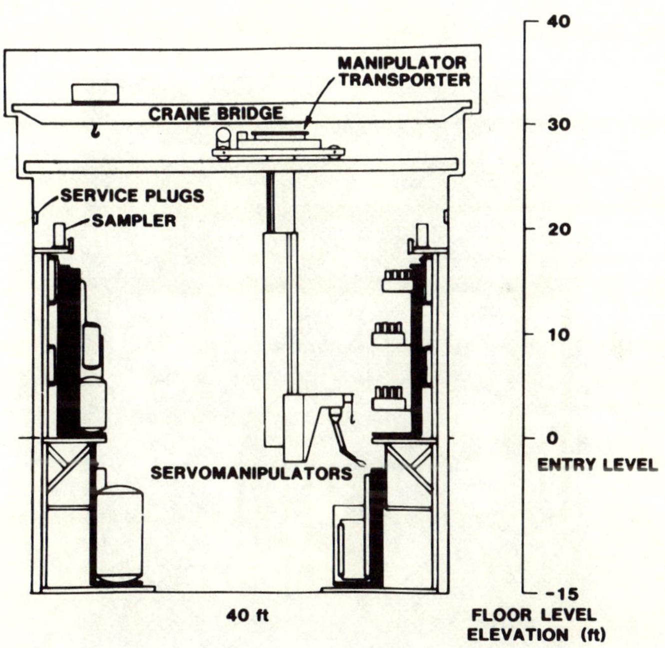

FIGURE 4 Cross section of main process cell of BRET.
From Oak Ridge National Laboratory

almost all of the reprocessing and waste management functions are integrated into a single large cell, which has entirely adequate space for the 100-kg/d throughput capacity plant. Only the product conversion and some of the peripheral waste handling would be done in adjacent areas. The Upper Process Cell immediately above the MPC will be used for transfer and special maintenance operations. The cross-section view shows the equipment modules on which all process equipment is installed. In the center aisle is space for movement of the bridge-mounted servomanipulator, which provides great versatility in replacing failed components on equipment racks or in disconnecting complete racks on rare occasions when major changes or repairs must be made. A maintenance test on a centrifugal contactor bank is shown in Fig. 5, and the remote control room from where the work was performed is shown in Fig. 6. (A motor was replaced in two hours.)

FIGURE 5 CRL model M-2 servomanipulator performing maintenance test on a centrifugal contactor unit.
From Oak Ridge National Laboratory

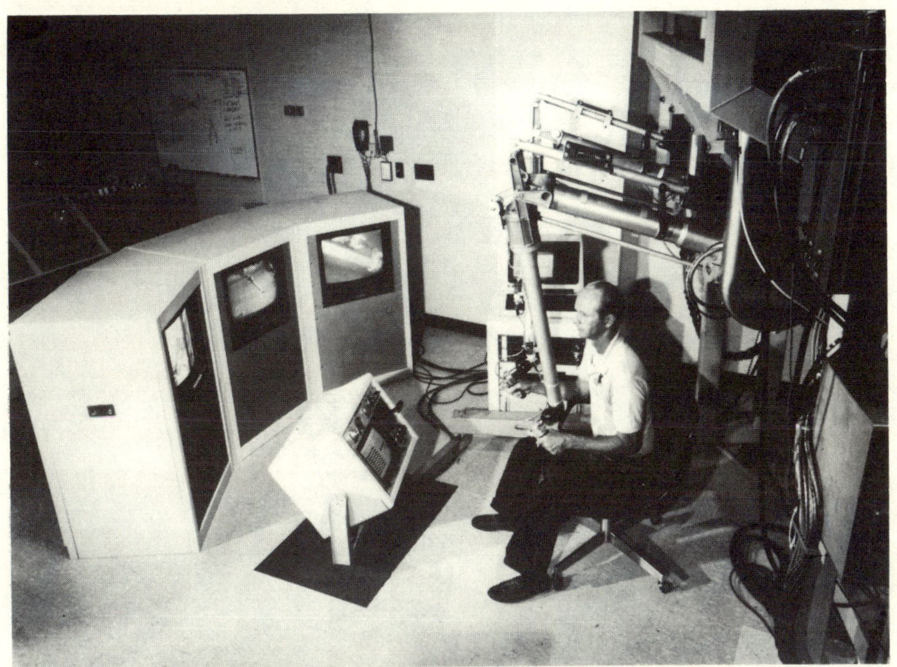

Fig. 6. Remote maintenance control room.
From Oak Ridge National Laboratory

The reliability and maintainability of the servomanipulator will be improved significantly with the Advanced Servomanipulator (ASM) that is now being installed in an area adjacent to the IET facility. The new manipulator, shown in Fig. 7, has undergone testing in a laboratory environment, using one of the arms as a pseudo-master in the early tests. The units have about a 1-lb sensitivity in the force feedback feature and a handling capacity of 50 lb. By using torque tubes and gear drives instead of the conventional tapes and cables, much higher reliability can be achieved, and the modular design permits replacement of failed modules with another manipulator.

While the long-range goals and missions of the breeder program are still unresolved, the ongoing development activities are reaching an advanced stage, bringing the new technology one step closer to reality. We expect to complete these development programs over the next few years.

FIGURE 7 Advanced servomanipulator on test stand.
From Oak Ridge National Laboratory

UK Experience in Fast Reactor Fuel Reprocessing

A.D. Frew and D.W. Harris, United Kingdom Atomic Energy Authority, Dounreay, Scotland

ABSTRACT

A decision in 1972 to reprocess the Prototype Fast reactor fuel at Dounreay maintained the close association of the establishment in the reprocessing field and was part of the evolutionary fast reactor fuel cycle strategy leading towards the design construction and operation of reprocessing plants to support demonstration fast reactors and subsequently a series of commercial fast reactors.

Considerable experience and expertise is being built up in the UK on the fast reactor fuel cycle through the operation of the small scale plant at Dounreay and has justified and endorsed the need for plant experience at a scale between laboratory developments and industrial plant.

This paper outlines the Dounreay plant operational experiences and associated development activities.

* Mr J D Frew is the Manager of the Reprocessing Operations Group at Dounreay

** Mr D W Harris is the Manager of the Reprocessing Development Group at Dounreay

INTRODUCTION

The history of fast reactor fuel reprocessing in the UK has been very thoroughly reviewed by R H Allardice (1)(2) since its inception in 1960 when irradiated fuel from the Dounreay fast Reactor (DFR) was successfully reprocessed and returned to the fuel cycle.

More recently following a decision to refurbish the DFR fuel reprocessing plant to support the Prototype Fast Reactor (PFR) fuel cycle (3) requiring 10^3 increase in plutonium throughput and a capability of accommodating a ten fold increase in fission product activity the reprocessing of PFR irradiated fuel has continued routinely since 1980 and to date more the 5.5 Te of mixed plutonium-uranium oxide fuel have been processed containing more than 1.1 Te of plutonium.

The operation of the plant against a "base load" of PFR irradiated fuel has provided a considerable amount of information on process and equipment technology and also has allowed specific investigations and experiments to be mounted on a plant scale and under realistic conditions aimed at improving the efficiency of the existing plant and also to make an essential contribution to conceptual and comparative cost studies which in turn guide the development programmes and provide a firm base for the design and operation of an industrial scale demonstration plant.

Routine and investigatory plant operations and an assessment of the data arisings have identified a number of improvements to the flowsheet and equipment and have increased our knowledge and understanding of the process chemistry and the effects and consequences of a hostile environment or plant and equipment. These could only have been obtained under plant scale conditions, thus confirming the evolutionary strategy in obtaining experience at the small plant scale before proceeding to commercial demonstration sizes and throughputs.

This paper reviews the experiences from the Dounreay PFR fuel reprocessing operation since 1980.

THE PROCESS

The process employed in the plant has been previously described (2)(4) and is illustrated in fig 1 briefly:-

The plant was designed to handle 180 day cooled core fuel containing recycled plutonium with a peak burn-up of about 10.0%. The calculated heat output of such a sub-assembly is about 3 kW.

Irradiated fuel is cleaned to remove sodium at the reactor, and the core plus integral axial breeder section of the sub-assembly is sent in a sealed transport container within a shielded flask to the reprocessing plant.

At the reprocessing plant a remotely operated laser is used to cut away part of the hexagonal sub-assembly wrapper to allow access to the fuel pins which are individually drawn. The pins are mechanically cropped into 5 cm lengths, collected in a basket and loaded into a batch dissolver. Nitric acid is used to dissolve the fuel and any insoluble materials are removed prior to solvent extraction process using a high-speed remotely operated centrifuge. The "insolubles" are stored in the centrifuge bowls.

The solvent extraction process is based on a three-cycle 20% tributyl phosphate and odourless kerosene (TBP/OK) flow sheet employing 100 mm equivalent 'diameter' geometry mixer settlers. The first cycle is used to extract the plutonium and uranium from the bulk of the fission products and the product is fed to the second cycle for further decontamination and separation of the U and Pu.

Equipment and layout constraints imposed by the original reprocessing plant precluded the use of a conventional oxidation-reduction flowsheet to separate plutonium from uranium. Sulphuric acid is added to separate uranium from plutonium by converting the plutonium to the sulphate which is not soluble in the organic phase. The uranium present is converted to the relatively insoluble sulphate but it is possible by controlling the nitric acid/sulphuric acid ratio in the process, to obtain the required separation, retaining the uranium in the organic phase.

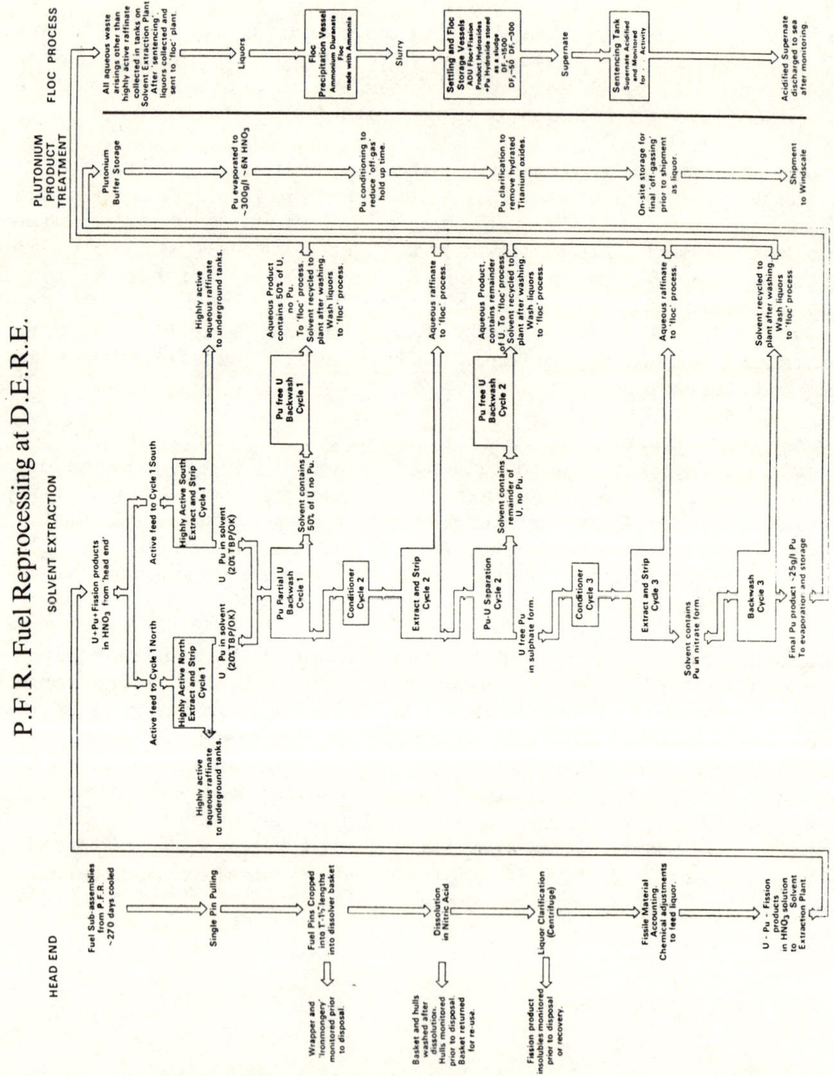

P.F.R. Fuel Reprocessing at D.E.R.E.

In the third cycle, the plutonium product is given a final purification before being concentrated by evaporation for transportation as nitrate to the fuel fabrication plant at BNFL, Sellafield for reuse in the PFR fuel sub assembly and irradiation in PFR thus completing the Fuel Cycle.

Advantage is taken of the availability of uranium from the process to treat the medium and low active raffinate by a uranic - ammonia "floc" process to reduce actinide and fission products to safe levels prior to sea discharge.

PLANT EXPERIENCE AND ACHIEVEMENTS

The plant has been operated in campaigns, each of which may have included differing feeds, requiring the process flowsheet to be altered to meet the specific requirements. Advantage has been taken of the intervals between campaigns to carry out engineering modifications and improvements to increase the reliability and performance of the plant and also to make provision for experimental equipment or flowsheets to be tested under plant scale conditions.

Since 1980 there have been (to January 1985) four major operating campaigns and more than 5.5 Te of mixed oxide has been processed containing more than 1.1 Te plutonium. Comparative details between actual plant performance and design parameters are shown in Table 1.

TABLE I PFR Fuel Reprocessing Plant Performance

	Achieved to Date	Achievement Relative to Typical PFR Fuel	Achievement Relative to Typical CFR Fuel
Peak burn-up	8.3%	0.99	0.93
Final heat rating in core	2.3 MW	0.33	0.26
Decay heat to reprocessing plant	1.2 kW	0.56	0.33
Cooling time	136 days	1.32	1.32
Fission Products (TBq per sub-assembly)			
Zr-95	555	0.54	0.28
Nb-95	1000	0.50	0.26
Ru-103	370	1.0	0.45
Ru-106	1440	0.46	0.31
Ce-144	1180	0.42	0.26
Total Beta	10400	0.55	0.33

Significant plant operational and management experience has been obtained demonstrating the feasibility of the process and together with the information derived from the research and development programmes provides a firm base and a continuing contribution to the design concept studies related to a larger scale plant.

Experiences of the various aspects of the process are now considered in more detail.

SODIUM CLEANING

Sodium removal from the PFR sub-assemblies is undertaken at the reactor using a steam inert gas mixture followed by a demineralised water wash and has proved to be a safe controllable and technically successful operation. The principle has been adopted for CFR fuel reprocessing and development work is aimed at reducing the cycle-time, hence the number of units required and also at the cleaning of sub-assemblies having shorter cooling times and correspondingly higher decay heats than currently processed.

TRANSPORTATION AND STORAGE

The transport of the sodium cleaned PFR sub-assembly from the reactor to the reprocessing plant is within the Dounreay Site; the sodium cleaned sub-assemblies are stored in nitrogen filled canisters under water in a buffer store and are taken as required to the reprocessing plant in shielded flasks. This successfully met the logistics of the local requirements within the limitations of decay heat levels up to 1.2 kW.

Large scale off site transportation is the subject of a separate study in which the technical safety and cost aspects of transportation in sodium or in nitrogen are being considered.

An experiment to examine the concept of storing sodium cleaned bare sub-assemblies under water, showed on immersion, a rapid but low level increase in the water activity associated with Mn_{54} and Co_{60}; however over the two years storage period these activities, which are thought to have been due to the corrosion of the reactor-sodium sensitised fuel cladding, have decreased.

DISMANTLING AND PIN CROPPING

The laser cutting of the sub-assembly wrapper to allow access to the ends of the fuel pins has proved to be simple and reliable and has not adversely affected pin cladding or the removal of the pins through the grids. An experiment to examine the effects of an inadvertent impingement of a laser beam onto the fuel pin cladding showed that the breached pin could be processed and there was no effect on the solubility of the fuel pellets in nitric acid.

No problems have been met in the routine extraction of single pins from the sub-assembly; data on pin pulling loads have been obtained to support an alternative dismantling concept based on removing the wrapper and grids from a complete bundle of pins and a preliminary experiment on an irradiated sub-assembly has demonstrated its feasibility.

The single pin cropping of fuel pins into 2.5 or 5.0 cm lengths has been accomplished without problems, providing an acceptable dissolver feed which has not deteriorated with time. However, the cropping rate is slower than desirable and alternative concepts have and are being examined to increase the throughput which will also take into account the use of modular units to aid maintenance, replacement and/or disposal. The principles of a double pin shear have been demonstrated on irradiated PFR fuel pins which is also being developed for the CFR plant.

DISSOLUTION

The dissolution in the batch dissolver of the PFR irradiated fuel has been accomplished without difficulty, indeed there is evidence that the fuel dissolves more rapidly and completely than is predicted from laboratory experiments. The "insolubles", undissolved fuel or fission products, are associated either with the hulls or the residue in the centrifugal clarifier storage bowls. From the reprocessing of 5.5 Te of mixed oxide fuel, the estimated plutonium in hulls or bowls as % of plutonium charged is 0.2 and 0.07 respectively (Table 2).

Fuel containing up to 40% as plutonium-uranium mixed oxide has been successfully dissolved, requiring appropriate criticality control in the batch dissolver by the addition of cadmium to the dissolver solution.

A visual inspection of the inside surface of the dissolver charge limb using a TV camera showed no evidence of corrosion. An investigation into the behaviour of a number of alternative candidate dissolver materials includes the long term testing of coupon specimens in the plant dissolver.

DISSOLVER LIQUOR CLARIFICATION

The removal of the fission product and fuel insolubles from the dissolver liquor has been successfully demonstrated using a centrifuge capable of generating some 10,000 g removing about 70% of the greater than 1 u dia insolubles. It is inferred that this is an acceptable degree of liquor clarification since no problems which could be associated with insoluble particulates have been met in the solvent extraction processes.

Modifications to the centrifuge to allow more effective control and use of the storage bowls are being examined on the plant scale.

PLUTONIUM VALENCY CONDITIONING

The clarified dissolver liquor is conditioned in the accountancy tank with sodium nitrite to alter the plutonium valency from Pu VI to Pu IV and this process has worked well; but recently interesting observations have been made which suggest that the plutonium valency conditioning by sodium nitrite may be unnecessary providing an adequate residence time in the accountancy vessel can be obtained and also that sodium nitrite conditioning may be relatively ineffective with low concentrations of plutonium such as under plant wash out conditions. These phenomena are being investigated but no conclusions can yet be drawn.

SOLVENT EXTRACTION

The PFR fuel reprocessing flowsheet uses a mixture of sulphuric and nitric acids as a partitioning agent adopted as previously mentioned, because of equipment and layout constraints imposed by the original plant; it has proved to be an extremely flexible and tolerant flowsheet with low plutonium losses to the raffinates Table 2 and it has not given rise to any equipment corrosion problems.

Experiments have been carried out to eliminate or reduce the additions of "salts" largely sodium and iron, to the plant as specified in the original flowsheet. This would be beneficial in terms of reducing waste storage volumes such as the liquid raffinates and the "floc" from the treatment of the medium and low active raffinates. The experiments examined the effects of reducing sodium and iron salt additions and showed that the specific solvent wash treatment (sodium carbonate and sodium hydroxide) was about optimum and that hydroxylamine carbonate was an effective "salt free" alternative; also iron additions could be eliminated without prejudice resulting in a lower "floc" storage volume.

The solvent extraction equipment, mixer settlers, performed well, with fission product decontamination factors in cycle 1 of 4×10^4; cycle II of 50 and cycle III of 5 and overall of 10^7. Losses to effluent steams have been very low ie plutonium as % of plutonium charged, (Table 2) stored in the high active raffinate 0.02%; stored as "flocced" medium and low active raffinate 0.1% and 0.02% discharged to sea.

The plutonium product specification for refabrication into mixed (Pu U)O_2 fuel has been met with an overall efficiency of better than 99.5 %.

WASTES AND EMISSIONS

The policy for the disposal of nuclear wastes is broadly to segregate the wastes into clearly defined levels of activity which can be consigned to optimised storage or disposal sites. Currently the highly active raffinate is stored in tanks until evaporation and vitrification facilities are available, medium and low active raffinates are treated by a floc process; the very low activity aqueous raffinate is discharged to sea and the "floc" combining the actinides and fission products are stored. In order to reduce levels even further below the accepted discharge limits - ion exchange methods are being studied on a plant scale.

Gaseous emissions have been kept within the internationally accepted guidelines and recent experiments to examine gaseous and particulate activity arisings in the ventilation streams have confirmed the high efficiency of the HEPA filters. Some preliminary observations have indicated that particulates in the gaseous effluents can arise during fuel dissolution and liquor clarification in addition to the expected fuel pin cropping operations. Investigations are planned to examine in detail the sources and character of such particulates which will provide valuable information to support the design of ventilation systems, plant equipment layout and containment.

In order to optimise the storage of wrapper materials an ultrasonic cleaning unit to reduce the plutonium surface contamination and therefore the storage category from the plutonium contaminated beta gamma active to the cheaper beta gamma is being installed.

TABLE 2 Plutonium Utilisation Efficiency (as % of Pu input)

(a)	Irrecoverable Wastes		
	High active effluent	0.02	
	Discharged to sea	0.02	
	Dissolver residues	0.07	
		0.11	
(b)	Recoverable Wastes		
	Sub-assembly solid wastes	0.03	
	Fuel Cladding	0.2	
	Medium/low active effluent (floc)	0.1	
		0.33	
	TOTAL		0.44
	ie once through plutonium efficiency	99.5	

TECHNICAL SUPPORT

Plant control, instrumentation and data retrieval systems are being studied in the PFR fuel reprocessing plant to improve existing control methods and systems to provide an extended data base of information for more advanced control systems and also to provide information for the larger scale plant design. Experience to date has shown the desirability of better control of inactive process feeds; work is in hand to improve the performance of feed control systems together with the development of plant mathematical models for use with statistical methods of plant control.

The routine use of analytical chemistry for process control is expensive and other methods are being examined for plant control in addition to alternative on line physical methods of analysis.

GENERAL AND ENGINEERING

As one would expect from a plant with many novel features operation of some parts of the plant have not been without problems. The dismantling cave crane failed due to insulation breakdown in the electrical cabling caused by a combination of radiation damage and mechanical damage. The crane was dismantled and replaced.

Master slave manipulator protective gaiters have a short operating life and their failure has led to serious contamination of manipulators making them difficult to remove and repair. Gaiter improvements are in hand and shielded manipulator decontamination and repair facilities are being constructed.

Early operation revealed that the dissolver charge cell shield windows were too small for efficient operation of the cell and after the third campaign the cell was decontaminated remotely from 10^3 Sv/h to 0.25 Sv/h using high pressure water jets and this allowed the windows to be replaced and only low operator dose occurred.

Apart from the need to remedy level control weir failure in the first cycle mixer settlers and a blockage caused by construction debris in the second cycle solvent recycle line it has not been necessary to carry out any repair work in the solvent extraction plant.

Some minor pipework modifications are being made in the first cycle extractor cells to allow the connection of future facilities such as an additional dissolver and a raffinate steam strip column.

RADIATION DOSES TO OPERATORS

Radiation doses have been well controlled with no operators exceeding 0.015 Sv in any year and the average well below 0.01 Sv/y. About 50% of the total dose has come from intervention maintenance with only low doses being experienced during normal operation. Except during periods when containment has been broken for maintenance or repair contamination has been at or near natural background level. The intervention dose levels will be further reduced with the introduction of new maintenance facilities presently being constructed.

CONCLUSIONS

(a) The Dounreay reprocessing plant has met the requirements imposed by the need to reprocess the irradiated fuel from the Prototype Fast Reactor.

(b) The feasibility of reprocessing at high efficiency high burn-up high Plutonium content fuel after short decay periods has been demonstrated.

(c) The routine and experimental operation of the plant is providing valuable information justifying the strategic policy decision to proceed to large scale commercial demonstration plants via the PFR fuel cycle, including this small scale reprocessing plant.

(d) The information obtained is being used for the design of reprocessing plant to meet the requirements of commercial scale fast reactors, including plant availability and cost effective technology which will provide a contribution to the estimates of fuel cycle costs as required by the Electrical Utilities for decision taking on the introduction of a series fast reactor system.

ACKNOWLEDGEMENT

The authors gratefully acknowledge the encouragement and support given by the Director of Dounreay in reprocessing development and operation and for permission to publish this paper.

REFERENCES

1. Nuclear Fuel Reprocessing in the UK. Allardice R H, Harris D W, Mills A L, Nuclear Power Technology Editor W Marshall OUP 1967.

2. Experience Gained in Processing PFR Irradiated Fuel - BNES Conference London 1980. Allardice, R H et al.

3. Barrett T R and Thom D. The Decommissioning and Reconstruction of the Fast Reactor Fuel Reprocessing Plant. IAEA SM234/9. IAEA Vienna 1978.

4. Mills A L. PFR Fuel Reprocessing Scheme Flowsheets and Chemistry Fast Reactor Fuel reprocessing. Society of Chemical Industry. May 1979.

SOLVENT EXTRACTION AND ION EXCHANGE IN RADIOACTIVE WASTE TREATMENT

Survey of Solvent Extraction and Ion Exchange in Radioactive Waste Processing

K.W. Carley-Macauly, Chemical Technology Division, AERE, Harwell, Oxon

THE WASTE TO BE TREATED

The fuel cycle consists of the following four broad stages, viz:

(i) Uranium extraction and purification.
(ii) Fuel Manufacture.
(iii) Reactor operations.
(iv) Fuel reprocessing.

The product from reprocessing is of course recycled to fuel manufacture.

Radioactive liquid wastes arise at all these stages of the fuel cycle. The largest potential arisings are from the wet chemical processes, i.e. (i) and (iv), but the other stages also give decontamination liquids, analytical residues, fuel recovery solutions, fuel pond storage, and other contaminated wastes from reactors. However efficient the individual process may be at recycling liquids, there is always a need for some discharge from the operating site as a whole, for instance to prevent the build up of inactive chemicals and also because of the very large volumes of water that may be involved in cooling operations, for example. Thus in general we can distinguish between treatment of individual process wastes and that of general effluents: this paper deals mainly with the latter.

It is worth noting that most (over 90%) of the radioactivity generated in the entire fuel cycle, arises in concentrated form in the high level waste (first cycle raffinates) from reprocessing operations.

OBJECTIVES AND PRINCIPLES

The general aim of waste treatment, for medium and low level liquids at least, is

(a) To reduce the activity in the bulk liquid to a level where that liquid is suitable for discharge or re-use. Suitability here includes meeting any regulatory requirements.

(b) To concentrate the active species removed from the bulk liquid into a suitable form for storage and eventual disposal. Among other requirements, this often implies that the volume should be as low as practicable because of the costs of storage and disposal.

(c) In particular cases there may be an additional requirement to separate forms of activity in order to facilitate storage and disposal.

(d) To design and operate the processes in the most economic way.

Among other effects, this last criterion often favours the use of a centralised plant, either for the treatment of blended effluents or for use with separate arisings on a jobbing basis.

An ion exchange column (or other device) will give a decontamination factor (DF) for the radioactive species of interest, which depends on the partition coefficient between liquid and solid, and also on the kinetics of exchange, the flow pattern of the liquid, and other factors. Typically, a DF of 100 or greater will be required of a useful ion exchanger. The capacity of the column (i.e. the bed volumes (B) of liquid before breakthrough) depends on the capacity of the exchanger itself and on the quantity of competing ions. Hence, ion exchange is most useful for decontamination of waters with low total solutes (< 1 g/ℓ say), or where the activity can be removed selectively. If the bed is used on a once-through basis, the concentration of activity is simply equal to (B) but it may be reduced somewhat by the subsequent immobilisation of the resin before disposal. If the column is regenerated, then what matters is the amount of salts in the regeneration liquid after neutralisation, since this will determine the limits for subsequent concentration by evaporation, and the volume of final immobilised solid. Typically, the volume reduction factors range from several hundred for "difficult" liquids, to many thousands.

Solvent extraction can be even more specific than ion exchange for particular species, and can give very high decontamination factors by virtue of true counter-current operation. One limit is the presence of impurities produced by radiolysis in the extractant. With a conventional solvent extraction flowsheet, however, the primary volume reduction factor is limited to the ratio between the feed volume and that of the strip solution. Since this ratio seldom exceeds a factor of 10, the process is of rather limited use for treating low activity waters, although it can have uses, for more active wastes, in the separation of activity as a prelude to further treatment.

KNOWN APPLICATIONS OF PROCESSES

General

Two recent reviews [1,2] on the treatment of medium and low active waste contain useful sections on ion exchange, but little or no mention of solvent extraction, and this may reflect the lack so far of established applications for the latter. High level liquid waste, by contrast, generally emerges directly from a solvent extraction process and further solvent extraction is being considered as a possible stage in its longer term management. In the sections which follow the applications, primarily of ion exchange, are discussed in relation to the separate stages of the nuclear fuel cycle.

Nuclear Power Plants

The wide use of ion exchange for the treatment of water at nuclear power stations may well have been inspired by its more general use for boiler water treatment. Thus ion exchange is used in water reactors to maintain the purity of coolants and to control the neutron absorbent properties of the moderator and hence the reactivity of the reactor itself. An interesting feature here is the use of powdered resins, as a filter precoat in place of the more conventional use of a filter followed by a packed bed of ion exchanger. This will be particularly encouraged where there is already a preference for once-through operation rather than regeneration of the ion exchanger. For instance, a description is given [1] for treatment of liquid wastes at a Finnish PWR station, and the current waste treatment plants for CEGB power stations provide cation and anion ion exchange units for standby use if required: here as elsewhere the wastes are segregated according to type and monitored on a batch basis. By contrast the French have generally used evaporation as a major method of treatment for active wastes at their PWR stations, but have shown in recent investigations [3] that the volume of solid waste can be reduced eleven-fold by using ion exchange resins to treat the same effluent and it has now been decided to install filters and ion exchange columns at all their PWR stations. This point illustrates the selectivity of ion exchange: with evaporation the whole burden of original dissolved salts has to be immobilised; with ion exchange by contrast the amount of salts in the regenerant solution should be very much less and only that, together with the eventually exhausted resins, has to be immobilised. Alternatively, the resin may be used on a once-through basis but that again generally shows an appreciable advantage.

In other cases, where a very high DF is needed, evaporation may be followed by an ion exchange treatment of the condensate, to remove the last traces of activity.

One source of active water at all power stations is the storage pond for irradiated fuel elements. Here the activity to be removed will consist of activation products from the metal cladding and/or mobile fission products, particularly caesium, from any small leaks in the fuel elements. At the same time, if the water is to be recirculated the chemical condition must be carefully maintained to avoid corrosion of the element cladding. Workers from CEGB [4] have described a treatment sequence at the Trawsfynydd Power Station to permit the purification of the recycled alkaline water together with the treatment of a lead stream destined for discharge. In this process all the activity is eventually removed in the adsorbed form on ion exchangers so that no evaporation of effluents is required.

The final stage of waste treatment in this case will involve the encapsulation of the ion exchanger in a water extendable resin in concrete containers, to await disposal [5].

The main alternative for the treatment of spent active resins is encapsulation in concrete. This process may require pre-conditioning of the resin to avoid displacement of activity by the high concentration of Ca in concrete water.

Reprocessing, Fuel and Development Plants

A similar application to that last mentioned is in the treatment of the pond water for storage of Magnox fuels at BNF Sellafield [1,6]. This pond is run on a once-through basis and the treatment consists of a single circuit of filtration and two ion exchange beds in series. The relative simplicity is partly dictated by the very large volumes to be treated - up to 4000 m^3/day - but is also facilitated by the excellent properties of the ion exchanger used. This is a natural zeolite, clinoptilolite, which provides a DF of over 100 for (chemically) trace quantities of Cs and Sr in the presence of 100 ppm Na. Up to 40,000 bed volumes can be processed before the leading ion exchange bed needs to be discharged.

Many nuclear undertakings, including R and D establishments, produce site effluents containing a wide range of radionuclides, often in variable quantities. In general, some form of chemical precipitation of floc treatment has been preferred for the clean up of such wastes, but ion exchange may then be used to "polish" the effluent by removing one or more particular radionuclides which are insufficiently removed by the main treatment. One example is at the French reprocessing plant of Cap la Hague which uses precipitation to remove alpha activity and Cs (by nickel ferrocyanide) but then passes the stream through a carboxylic exchanger in sodium form for the removal of strontium. The regeneration liquors from this ion exchange operations pass to the Medium Active treatment line, where Sr is finally removed by coprecipitation with barium sulphate [2,11]. Somewhat similarly the Russians back up a ferric hydroxide precipitation of Research Institute wastes, with cation and anion exchange columns [1].

Several examples of this use of ion exchangers for polishing a treated waste are also available for the United States [2,7] where natural or artificial zeolites are used at a number of plants to remove particularly Cs. For the actinides, which may be in combination with organic species, bone char has been used at the Mound Laboratories [8] and active carbon has also been proposed. The use of preformed ferrites has been studied at Rocky Flats, also for the removal of actinides, and as an alternative to in-situ precipitation of the same materials [9,10].

Packaged Ion Exchangers

Where ion exchangers are to be used on a once-through basis, and particularly for localised or temporary applications, there may be particular attractions in the use of a container which serves both for the process and the ultimate disposal as this can obviate the need for elaborate facilities for the removal and encapsulation of the ion exchange granules. Examples are the EPICOR-II system, used for the treatment of water from the accident at the Three Mile Island Power Station, and the use of skips containing zeolite in the Windscale fuel pond pending the completion of the SIXEP plant mentioned above [1]. This principle may have further future applications in the treatment of wastes from individual plants, particularly where these are intermittent or would give problems in transport.

Uranium Ore Processing

This area is often regarded as remote, perhaps because of its geographical location as much as of the distinctive nature of the operations, which basically involve only natural radioactivity. Such activity while perfectly safe at least by the standards of our planet, in its native rock deposits, may cause concern if distributed into local streams. The species of concern here are mainly ^{230}Th and Ra as radioactive species, and Pb on grounds of chemical toxicity. Current treatment processes are based on precipitation, e.g. of thorium at high pH and of radium by coprecipitation with $BaSO_4$, but a recent report [11] points the way to the use of ion exchangers. Thorium may be removed as a cation or an anion (e.g. in nitrate complex form) or even, it seems, in the form of a complex with EDTA or citrate, and radium (which only occurs in a divalent form) by cation exchange. Indeed, the latter operation is already carried out in other places for the treatment of naturally radioactive waters to provide drinking water. The review cited also concludes that solvent extraction would have little use in treating these mine waters.

Highly Active Waste Treatment

The first cycle raffinates from solvent extraction treatment of reprocessing liquors contain virtually all the fission products together with the trivalent actinides (Am, Cm) and small quantities of uranium, plutonium and neptunium. Before storage these raffinates are evaporated close to the point of incipient precipitation, and the distillate may be further purified as by ion exchange. The next objective is the solidification in a form suitable for ultimate disposal. While it is generally proposed to vitrify the highly active waste directly, research has also been carried out on the use of inorganic exchangers such as sodium titanate to demineralise this waste, i.e. to absorb all the fission products [1]. In fact this inorganic exchanger can give DFs of up to 10^8 for most fission products, but a final passage through a zeolite bed is needed to complete the removal of Cs. The attraction of such processes is that the exchangers can be subsequently heat treated to produce a massive insoluble waste form comparable with that from the more established vitrification processes.

Also in the research phase are various schemes to partition the highly active waste. The purpose here would be to separate the long lived actinide species from the more intensely active, but shorter lived, fission products so as to allow their separate storage and/or disposal. (There is considerable debate about whether or not this is a worthwhile step). The separation is complicated by the overlap in chemistry, particularly between the trivalent actinides and the lanthanides. Although ion exchange has been used elsewhere in reprocessing operations for the separation and purification of Pu in particular [10] the greater separative power of solvent extraction would be at an advantage for the complete separation of all the actinides. Even so, relatively exotic extractants [12] and complicated flowsheets [13,14,15] are needed for this purpose. As an example, one of three processes suggested by workers at JRC, Ispra [13], requires the denitration of highly active waste to about 0.1 M HNO_3 followed by the extraction of rare earths and actinides into HDEHP (bis-2-ethylhexyl phosphoric acid). After this, Am and Cm are stripped into DTPA (di-ethylene triamine pentacetic acid), the rare earths are stripped into nitric acid, and any remaining Pu and Np into oxalic acid. Other studies in this area [16] have also considered the use of extraction chromatography in which the solvent DHDECMP was sorbed on an "Amberlite" resin. This technique can be used to combine the extractive power of a solvent, with the relative simplicity of operation of an ion exchange column.

FUTURE PROSPECTS AND DEVELOPMENTS

The previous section on highly active wastes has strayed well into the area of research and development, an area also where the justification may be based as much on perceived political needs as on pure technical requirement. Turning back to medium and low level wastes, there is a continuing interest in more selective ion exchangers, to reduce the final volume of the active solids. Perhaps the ultimate in this direction is claimed by the Devoe-Holbein Company for their new compositions who structure is modelled on biological mechanisms of metal extraction, and which can be quite specific for individual metals. These compositions are regenerable using only one or two bed volumes of eluent, which again would confer a considerable advantage provided the initial capacity were suitably high. (For instance, the volume reductions of 250-1500 (i.e. feed volume to eluent volume) reported [17] for treatment of fuel pond storage water have to be compared with factors of $20-50 \times 10^3$ for particular zeolites (feed:bed volume) used on a once-through basis. This means that the eluent in the first case might then need to be reduced 10-100 fold by evaporation (if possible) to give a comparable final volume).

There is also a need for sorbents capable of removing the actinides which in neutral or alkaline waters are likely to be present as polymeric hydroxides or in various complexed forms at low concentrations. Section 3.1 has already mentioned removal of citrate or EDTA complexes of thorium, and there is also hope that inorganic ion exchangers may prove useful, either by (anionic) sorption of the complete complex group or by removing the actinide itself from the complex, as in the case of uranyl group absorption from a complexed anion onto hydrous titanium oxide [18]. Inorganic exchangers are often of an amphoteric nature and can be highly specific for individual species under particular conditions. They have an advantage for conditioning to stable inorganic forms for disposal, perhaps even by hot pressing to yield a ceramic form. One disadvantage had been their relative fragility, compared to the robust organic beads, but it is now proving possible to manufacture them as spheres by a gel precipitation technique [19].

A special area where inorganic exchangers have been considered is in the treatment of solvent from the main reprocessing separation process. The TBP hydrolyses to some extent under radiation to form hydrogen dibutylphosphate which tends to extract some fission products into the solvent phase and thus needs to be removed. Normally this is done by a caustic/carbonate wash process but this in turn is a major source of salts in the reprocessing waste. A direct treatment process to yield an inorganic solid would thus be attractive, and it has been shown that hydrous titanium oxide can perform this function [20], although so far perhaps not with sufficient capacity to make the process really attractive for remote operation.

A further possibility currently being explored for aqueous treatment is that of an electrochemically controlled ion exchanger [2,19] in which the conditions for absorption or elution of the active species are created by electrochemical generation of OH^- or H^+ in the region of the functional sites. By this technique the initial absorption can be enhanced and elution can take place with minimal amounts of added anion, or even into distilled water alone, thus reducing the amount of salt present in the final waste concentrate.

The above areas are under continuing study by the nuclear industry together with the application of existing and new organic ion exchangers.

As we have seen, solvent extraction in a conventional mode may not provide adequate concentration factors for treatment of dilute wastes but it could have application for the separation of actinide activity from other wastes, either for strategic reasons or because the inorganic species present cannot be separated by other means. However, other techniques of applying specific solvents which allow a greater concentration ratio may prove more fruitful, and these include liquid membrane techniques [1] and as sorbed species [2].

REFERENCES

[1] IAEA Technical Reports Series No. 236, Vienna, 1984.

[2] Carley-Macauly, K.W., Gutman, R.G. et al, eds. "Radioactive Waste Management, Vol.1: Advanced Management Methods for Medium Active Liquid Waste". Harwood Academic Publishers for CEC, 1981. ISSN: 0275-7273.

[3] Pottier, P. et al. IAEA-SR-57-11. Paper to IAEA Seminar, Karlsruhe, October 1981.

[4] Broadbent, D., Haighman, A.P. et al in Proc. Conf. on On-Site Management of Power Reactor Wastes, Zurich, OECD, 1979, p.97.

[5] Willis, A.S.D., Haigh, C.P. et al. In Radioactive Waste Management (Conference), BNES, London, 1984, p.89.

[6] Howden, M. In "Effluent Treatment in the Process Industries" (1983), Instn. of Chem. Engrs. Symp. Series No.77, p.253.

[7] IAEA Technical Reports Series No.27, Vienna, 1964.

[8] Roberts, R.C. and Herald, W.R. "Development of Ultrafiltration and Adsorbents", April-Sept 1979. MLM 2684 (1980)

[9] Boyd, T.E., Kochen, R.L. and Price, M.Y. RFP-3324.

[10] Carley-Macauly, K.W., Gutman, R.G. et al. AERE-R-11304.

[11] Itzkovitch, I.J. and Ritcey, G.M. CANMET Report 79-21 (1979).

[12] Shoun, R.R. and McDarell W.J. In Am. Chem. Soc. Symp. Series No.117, "Actinide Separations" (April 1979), eds. J D Navratil and W.W. Schulz, Section 6, pp. 71-87.

[13] Cécille, L., Dworschak, H. et al. op cit, Ref.12, Section 30, pp. 427-440.

[14] Tedder, D.W. et al. Nuclear Technology 53, 19 (1981).

[15] Liljenzin, J.O. et al. Sepn. Sci. and Techn. 15(4), p. 799 (1980).

[16] Martella, L.L. et al. In "Actinide Recovery from Waste and Low Grade Sources", eds. J D Navratil and W.W. Schulz, Harwood Academic Publishers, 1982, Chapter 2, pp. 27-40.

[17] Holbein, D.E., DeVoe, I.W. et al. "Devoe Holbein Technology .....", paper to Massachusetts Hazardous Waste Source Reduction Conference, October 1984.

[18] Maya, L. Radiochimica Acta 31 (1982) 147.

[19] Turner, A.D. et al. Second European Conference on Radioactive Waste Management and Disposal, Luxembourg, April 1985 - Paper II-2.

[20] Healy, T.V. IAEA-SM-207/23.

Solvent Extraction and Recovery of the Transuranic Elements from Waste Solutions using the Truex Process

E. Philip Horwitz, Chemistry Division, Argonne National Laboratory, Argonne, Illinois, 60439, USA
Wallace W. Schulz, Rockwell Hanford Operation, Richland, Washington, 99352, USA

High-level liquid waste is produced during the processing of irradiated nuclear fuel by the PUREX process. In some cases the treatment of metallurgical scrap to recover the plutonium values also generates a nitric acid waste solution. Both waste solutions contain sufficient concentrations of transuranic elements (mostly ^{241}Am) to require handling and disposal as a TRU waste. This paper describes a recently developed solvent extraction/recovery process called TRUEX (transuranium extraction) which is designed to reduce the TRU concentration in nitric waste solutions to < 100 nCi/g of disposed form [1,2]. (In the USA, non-TRU waste is defined as <100 nCi of TRU/g of disposed form.) The process utilizes PUREX process solvent (TBP in a normal paraffinic hydrocarbon or carbon tetrachloride) modified by a small concentration of octyl(phenyl)-N,N-diisobutylcarbamoylmethylphosphine oxide (abbrev. CMPO). The presence of CMPO enables the modified PUREX process solvent to extract trivalent actinides as well as tetra- and hexavalent actinides. A major feature of the TRUEX process is that it is applicable to waste solutions containing a wide range of nitric acid, salt, and fission product concentrations and at the same time is very compatible with existing liquid-liquid extraction technology as usually practiced in a fuel reprocessing plant.

To date the process has been tested on two different types of synthetic waste solutions. The first solution is a typical high-level nitric acid waste and the second a typical waste solution generated in metallurgical scrap processing. Compositions of these synthetic wastes are shown in Tables 1 and 2.

TABLE 1 Typical Composition of Acidic High-Level Waste

Component	Concentration, M	Component	Concentration, M
HNO_3	1.0	Rh	$2.5x10^{-4}$
		Pd	$8.8x10^{-4}$
Na	0.20	Ag	$3.7x10^{-5}$
Al	0.60	Cd	$5.0x10^{-5}$
Cr	0.015	Te	$2.9x10^{-4}$
Fe	0.15	Cs	$1.2x10^{-3}$
Ni	0.007	Ba	$8.1x10^{-4}$
F^-	0.15	La	$6.1x10^{-3}$
SO_4^{2-}	0.20	Ce	$1.2x10^{-3}$
NO_2^-	0.005	Pr	$5.7x10^{-4}$
		Nd	$1.9x10^{-3}$
Se	$4.2x10^{-5}$	Sm	$3.9x10^{-4}$
Rb	$2.6x10^{-4}$	Eu	$7.5x10^{-5}$
Sr	$6.2x10^{-4}$	Gd	$1.2x10^{-5}$
Y	$3.5x10^{-4}$		
Zr	$2.6x10^{-3}$	U	12-48 mg/L
Mo	$2.4x10^{-3}$	Np	25-50 mg/L
Tc	$1.0x10^{-5}$	Pu	0.2-0.4 mg/L
Ru	$9.9x10^{-4}$	Am	0.005-0.05 mg/L

TABLE 2 Typical Composition of Plutonium Scrap Waste

Component	Concentration, M	Component	Concentration, M
HNO_3	1.5	Mn	0.003
		Fe	0.03
F	0.09	Ni	$4x10^{-4}$
		Cu	$3x10^{-4}$
Be	$7x10^{-5}$	Zn	$6x10^{-4}$
Na	0.04	Pb	$5x10^{-4}$
Mg	0.06		
Al	0.43	U	0.7-700 mg/L
K	0.003	Pu	3-14 mg/L
Ca	0.06	Am	1-3 mg/L
Cr	$4x10^{-4}$		

Two different TRUEX process solvent compositions are employed depending on the type of waste solution to be processed. TRUEX solvent used for processing high-level liquid waste requires a normal paraffinic hydrocarbon (NPH) diluent with an average carbon chain length of thirteen. Such diluents have flash points of ~ 100°C which is well above the minimum of 80°C that is usually required for PUREX solvent used to process irradiated nuclear fuel. TRUEX process solvent containing 0.2 M CMPO - 1.4 M TBP in Conoco C_{12}-C_{14} has been found to have optimum distribution ratios for americium and high loading capacities without third phase formation (1). (Conoco C_{12}-C_{14} is a commercial NPH containing 12% C_{12}, 56% C_{13}, and 28% C_{14}.) TRUEX solvent used for processing metallurgical scrap waste requires a nonflammable diluent such as carbon tetrachloride or tetrachloroethylene (TCE). Optimum distribution ratios for americium have been obtained with 0.25 M CMPO - 0.75 M TBP in CCl_4 or TCE [3]. Third phase formation does not occur with CCl_4 or TCE diluents even when the solvent is saturated with Am(III) or Pu(IV).

Tables 3 and 4 show the distribution ratios obtained for the major constituents of each waste solution using the appropriate TRUEX process solvent. The phase ratios and temperatures selected for the measurements were chosen to simulate the conditions in the first extraction stage.

Figure 1 shows the generic TRUEX process flowsheet. The process is generic because it is applicable to solutions containing a wide range of nitric acid, fission product, and salt concentrations. For each type of waste solution, the TRUEX process is capable of reducing TRU concentration by many orders of magnitude.

Acidic high-level waste solutions require the addition of oxalic acid to the feed to suppress extraction of Zr and Mo by forming inextractable oxalato complexes. The oxalic acid concentration is generally in the range of 0.15 to 0.25 M depending on the concentration of Al in the waste solution. (High concentrations of Al require high concentrations of oxalic acid.) As the data in Table 3 predict, essentially all of the Am, Pu, and U and a major portion of the Np are extracted (without any oxidation state adjustment) into the TRUEX process solvent. Rare earth fission products and Tc are extracted to varying extents depending on the organic/aqueous phase ratio and the number of extraction

TABLE 3 Distribution ratios from an acid high-level waste (0.15 M $H_2C_2O_4$)
Organic Phase 0.2 M CMPO – 1.4 M TBP–Conoco C_{12}–C_{14}
O/A = 0.5, T = 40°C

Component	D	Component	D
HNO_3	0.3	Rh	$<10^{-1}$
		Pd	0.4
Na	$<10^{-2}$	Ag	$<10^{-1}$
Al	$<10^{-2}$	Cd	$<10^{-1}$
Cr	$<10^{-2}$	Te	$<10^{-1}$
Fe	0.05	Cs	$<10^{-2}$
Ni	$<10^{-2}$	Ba	$<10^{-2}$
F^-	–	La	4
SO_4^{2-}	–	Ce	6
NO_2^-	–	Pr	6
		Nd	6
Se	$<10^{-2}$	Sm	6
Rb	$<10^{-2}$	Eu	5
Sr	$<10^{-2}$	Gd	4
Y	2		
Zr	0.04	U(VI)	$>10^2$
Mo	0.2	Np(IV), Np(V)	$>10^3$, $\lesssim 0.5$
Tc	3	Pu(IV)	$>10^3$
Ru	0.4	Am(III)	7

TABLE 4 Distribution ratios from a Pu scrap waste solution
Organic Phase 0.25 M CMPO – 0.75 M TBP–CCl_4
O/A = 0.33, T = 30°C

Component	D	Component	D
HNO_3	0.2	Mn	<0.01
		Fe	0.06
F	–	Bu	<0.01
		Cu	<0.01
Be	<0.04	Zn	<0.01
Na	<0.01	Pb	<0.01
Mg	$<0..001$		
Al	<0.0001	U(VI)	$>10^2$
K	<0.01	Pu(IV)	$>10^3$
Ca	<0.01	Am(III)	17
Cr	<0.001		

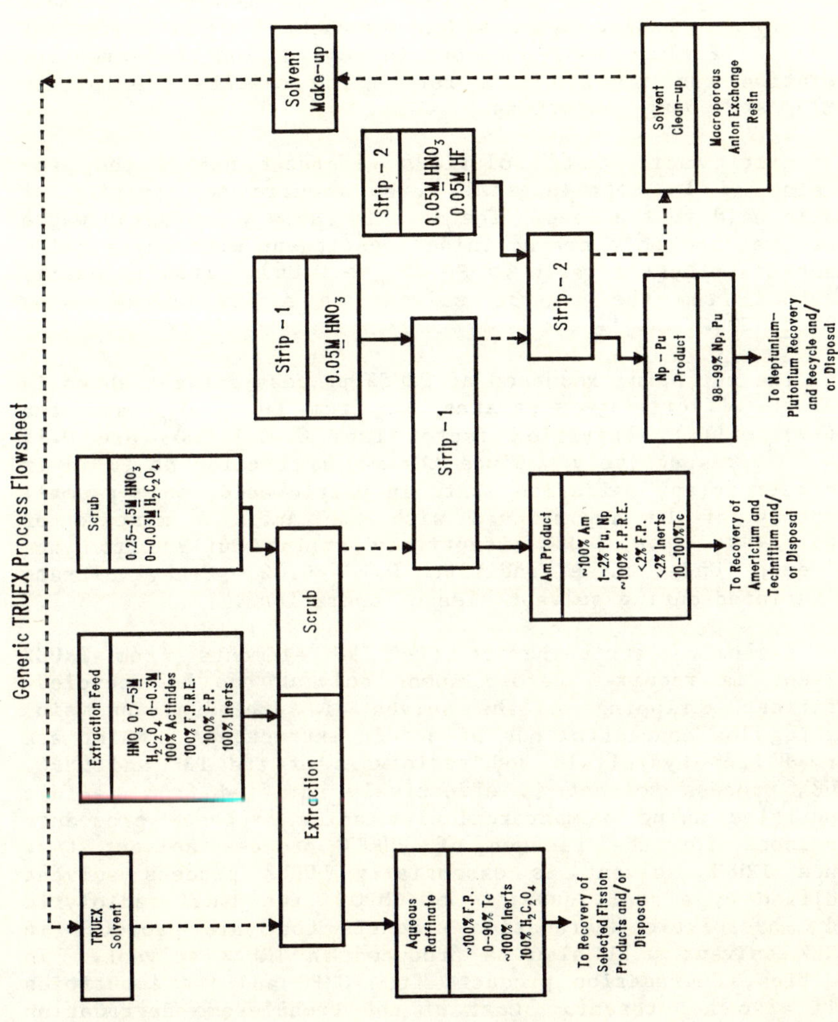
Generic TRUEX Process Flowsheet
TRUEX Solvent
Extraction Feed
HNO_3 0.7–5M
$H_2C_2O_4$ 0–0.3M
100% Actinides
100% F.P.R.E.
100% F.P.
100% Inerts
Scrub
0.25–1.5M HNO_3
0–0.03M $H_2C_2O_4$
Extraction
Scrub
Aqueous Raffinate
~100% F.P.
0–90% Tc
~100% Inerts
100% $H_2C_2O_4$
To Recovery of Selected Fission Products and/or Disposal
Strip – 1
0.05M HNO_3
Strip – 1
Am Product
~100% Am
1–2% Pu, Np
~100% F.P.R.E.
<2% F.P.
<2% Inerts
10–100% Tc
To Recovery of Americium and Technetium and/or Disposal
Strip – 2
0.05M HNO_3
0.05M HF
Strip – 2
Np – Pu Product
98–99% Np, Pu
To Neptunium–Plutonium Recovery and Recycle and/or Disposal
Solvent Make-up
Solvent Clean-up
Macroporous Anion Exchange Resin

FIGURE 1

stages. More than half of the Y and La may be rejected to the aqueous raffinate under proper conditions.

Extensive scrubbing of the organic phase is unnecessary since a few percent of non-rare earth fission products in the Am-rare earth product does not significantly increase the quantity of TRU waste. A small concentration of Fe(III), e.g. 0.005 M, in 0.5 M HNO_3 is used to remove the small quantity of oxalic acid which extracts into the organic solvent. Failure to remove oxalic acid during the scrubbing operation can result in a rare earth oxalate precipitate during stripping operations.

Nitric acid waste solutions generated during the processing of Pu metallurgical scrap require no addition of oxalic acid to the feed. The data in Table 4 for scrap waste show that the only non-actinide constituent with any significant distribution ratio is Fe (D_{Fe} = 0.06). Iron is easily scrubbed from the organic solvent in a few stages using 0.25 M HNO_3.

The stripping sequence of TRUEX process solvent shown in Fig. 1 effectively separates Am from Pu. D_{Am} and the Pu(IV)/Am(III) separation factor from 0.05 M HNO_3 are 0.15 and 10^3, respectively. Since the major fraction of Pu is in the tetravalent oxidation state in nitric acid, only a small quantity of Pu is removed with the Am. A mixture of 0.05 M HNO_3 - 0.05 M HF effectively strips Pu(IV) from the solvent. Under these conditions D_{Pu} = 0.02. Uranyl nitrate is stripped during solvent cleanup operations.

Extensive stripping of the TRU elements from TRUEX solvent is required before spent solvent can be recycled. Efficient stripping of the solvent is dependent on maintaining low concentrations of acidic extractants, which are formed from hydrolysis and radiolysis of the TBP and CMPO. TRUEX process solvent is effectively purified from acidic impurities using a macroreticular anion exchange procedure developed for the cleanup of PUREX process solvent (4). Since TRUEX solvent is essentially PUREX process solvent modified by a small quantity of CMPO, the usual radiolytic and hydrolytic degradation products that are produced in PUREX solvent will also be produced in TRUEX solvent. In addition, degradation products from CMPO and its impurities will also be present. Most of the troublesome degradation products from CMPO arise from impurities. Once these degradation products are removed with the macroreticular resin, TRUEX solvent can be adequately purified by the same Na_2CO_3 scrubbing procedures employed for PUREX process solvent.

Batch countercurrent test runs using a synthetic acidic high-level waste as feed has demonstrated the ability of the TRUEX process to effectively reduce the TRU concentration to < 100 nCi/g of solidified raffinate (2). Eight extraction and four scrub stages were used in the test. Decontamination factors of TRU elements from the waste solution were 5×10^4 (Am), $>>10^5$ (Pu and U), and 6 (Np). These decontamination factors reduced the total TRU activity in the raffinate to 40 nCi/g. The weight of TRU waste was reduced by 150 fold.

Batch extraction tests with synthetic Pu scrap waste reduced the Am and Pu concentrations by $>10^3$ and $>10^4$, respectively, in three extraction and two scrub stages. The process reduced the quantity of TRU waste by 10^3. Selective stripping of Am from Pu from the solvent was also demonstrated.

Continuous countercurrent extraction test runs using the ANL annular centrifugal contactors and synthetic Pu scrap waste are in progress at Argonne National Laboratory. Extensive hydrolytic and radiolytic stability and solvent cleanup testing is also in progress at Argonne.

Listed below is a summary of the major features of the TRUEX process.

1. TRUEX and PUREX have very similar extraction, scrub, strip, and solvent cleanup cycles.
2. TRUEX and PUREX process solvents have similar densities, interfacial tensions, and hydrolytic and radiolytic stabilities.
3. TRUEX process solvent can be prepared from existing PUREX solvent inventory since the quantity of CMPO required is small.
4. TRUEX process solvent will effectively extract actinides in the III, IV, and VI oxidation states; therefore, no oxidation state adjustment is necessary and high decontamination factors of TRU elements from waste are easily achieved.
5. TRUEX requires no acidity adjustment since D_{Am} is not very sensitive to nitric acid concentration in the range of 0.7 to 7 $\underline{M}$; therefore, the process is easy to control.
6. TRUEX does not increase the volume of solid waste or create difficult to handle waste streams.

ACKNOWLEDGEMENT

Funding for the development of the TRUEX process was provided by the Office of Basic Energy Sciences, Division of Chemical Sciences, U. S. Department of Energy under contract W-31-109-ENG-38 and the Rockwell Hanford Operation.

REFERENCES

1. E. P. Horwitz, D. G. Kalina, H. Diamond, G. F. Vandegrift, and W. W. Schulz, Solvent Extr. Ion Exch. 3, No. 1 (1985).
2. G. F. Vandegrift, R. A. Leonard, M. J. Steindler, E. P. Horwitz, L. J. Basile, H. Diamond, D. G. Kalina, and L. Kaplan, Report ANL-84-45, U. S. Department of Energy, Argonne, Illinois, 1984.
3. D. G. Kalina and E. P. Horwitz, Solvent Extr. Ion Exch. 3, No. 2 (1985).
4. W. W. Schulz, in Proceedings of the International Solvent Extraction Conference, Vol. 1, Society of Chemical Industry, London, p. 174 (1971).

The Application of Inorganic Ion Exchangers to the Treatment of Medium Active Effluents

E.W. Hooper, Atomic Energy Research Establishment, Harwell, Oxon

INTRODUCTION

In general, inorganic ion exchangers possess many properties which make them more suitable than organic exchange resins for the treatment of radioactive waste solutions, e.g.

- Many exhibit little or no change in chemical and physical properties when subjected to high radiation doses. This property is desirable if the loaded absorber is either to be regenerated or to provide a stable matrix for long term storage.

- Many inorganic exchangers do not suffer chemical degradation in the presence of HNO_3 whereas organic resins may form explosively unstable compounds.

- They can exhibit a high selectivity for certain ions. For example, ion exchange can provide a means for separating radioactive ions from inactive ions such as Na^+, NH_4^+, Fe^{3+}, Al^{3+} and Mg^{2+} which may be deleterious in final waste forms.

- Unlike organic resins, the functional group is an integral part of the material and as such both structure and exchange properties are governed by the same chemical composition.

- Inorganic exchangers are more compatible than organic resins with the hydrated cement-based matrices which are likely to be preferred for waste immobilisation.

- Further, in particular instances, hot-pressing the loaded inorganic absorber to give a dense, stable waste form may be a practicable option.

In 1982 the UK Department of the Environment initiated at AERE Harwell a programme of work with the objective of developing processes which would provide efficient and safe concentration of the radioactive consituents of waste streams into solid materials which can be consolidated into forms suitable for storage or disposal.

This paper summarises the data obtained so far on the absorption of plutonium, americium and neptunium from spiked solutions and of total α-activity and fission product removal from solutions of irradiated fuel.

REVIEW OF INORGANIC ION EXCHANGE MATERIALS

A detailed literature survey[1] has been carried out to collect together available information on the use of inorganic exchangers for the absorption and separation of radioactive nuclides. Information on precipitation processes used for radioactive waste treatment has also been collected, since the absorption of elements by precipitates is often by an ion-exchange process, and many inorganic ion exchangers are prepared by a process involving precipitation.

The materials of interest may be broadly classified as: natural inorganic exchangers (clay minerals and zeolites), synthetic zeolites, oxides and hydrous oxides, acidic salts of multivalent metals and cyanoferrates. Their useful features are outlined below:

Natural Inorganic Materials

The clay minerals and naturally occuring zeolites, though replaced to a large extent by synthetic exchangers, continue to be used and have been applied extensively to the treatment of radioactive waste solutions. Limitations to their use include:

(i) relatively low ion exchange capacities,

(ii) low abrasion resistance of the zeolites,

(iii) clay minerals tend to peptize, i.e. to form colloids in some solutions,

(iv) these materials are partially decomposed by acids and alkalis.

Synthetic Zeolites

These materials are commercially available; they act as molecular sieves by excluding ions larger than the opening in the crystalline matrix. Limitations to the use of synthetic zeolites are similar to those listed above for naturally occurring zeolites but the synthetic zeolites possess improved stability and a higher cation exchange capacity.

Oxides and Hydrous Oxides

Only a few of these materials are available in commercial quantities. Many show high exchange capacity, low solubility and good radiation and thermal stability.

Cation or anion exchange sorption by oxides and hydrous oxides is known to occur predominantly by displacement of hydrogen and hydroxide ions from the sorbents provided that these have previously been washed free from impurity ions. The predominant functional groups are probably metal-hydroxide bonds which can exhibit amphoteric properties favouring anion exchange at low pH and cation exchange at high pH.

Although hydrous oxides are at their most active when freshly precipitated, they are at their least useful as regards ion exchange column applications. Their relatively high ion exchange capacities are offset by their ease of dissolution in acids or bases and the poor filtration characteristics of the precipitates in general render them unsuitable for use as column sorbents.

If gelatinous precipitates of the hydrous oxides are dried they undergo considerable shrinkage to give glassy gels which break down in water to produce granular particles suitable for column applications. The structure and ion-exchange properties of these dried gels depend very much on the conditions of preparation of the precipitates and the drying procedure used to produce the granular product. During the drying stage it is probable that, apart from the removal of free water, elimination of water from -OH groups occurs to produce -M-O-M- bonds (M = metal) giving a stronger absorber material but reducing the ion-exchange capacity.

The quantitative effect of heat treatment of different oxides and hydrous oxides on their ion exchange properties varies considerably. Generally ion exchange capacities show a steady decrease with heating and at about 200°C have fallen to 20-30% of their values for low temperature dried material. Heat treatment can improve both the selectivity and the chemical stability of these absorbers.

The use of an oxide or hydrous oxide absorber as a column material is obviously governed to a large extent by its resistance towards breakdown or dissolution when treated with various reagents. In general the insoluble acidic oxides and hydrous oxides are stable in acid solutions but dissolve readily at pH > 7. In contrast the more basic hydrous oxides such as TiO_2 and ThO_2 dissolve readily in mineral acids above about 0.1M but are stable in less acid solutions.

Acidic Salts of Multivalent Metals

A wide range of compounds of this type have been described as ion-exchangers. These salts, acting mostly as cation exchangers, are gel-like or micro-crystalline materials with the composition and properties depending on the method of preparation.

In general they are not available in commercial quantities and many are likely to be expensive to produce because of the limited availability and high cost of the materials required in their preparation.

Cyanoferrates

The ion-exchange properties of a large number of insoluble cyanoferrates of various metals have been studied. These absorbers act as cation exchangers with an especially high affinity for caesium. The exchange capacity depends strongly on the method of preparation.

The cyanoferrate ion-exchangers are stable in mineral acids; in contact with strong HNO_3 a slight oxidation of FeII to FeIII can occur. In alkaline solutions a tendency to peptize has been observed but this can be greatly reduced by addition of neutral salts.

Irradiation of potassium cobalt hexacyanoferrate to a total dose of 6×10^8 rads produced slight decomposition.

Attempts to elute caesium absorbed on cyanoferrates have been only partially successful and in general is accompanied by decomposition of the absorber.

Cyanoferrates are currently available in research quantities but because of the relatively simple method of preparation it is thought that larger scale manufacture would be feasible and economic.

An assessment exercise[1], based on the literature search, identified six ion exchangers with a known affinity for actinides and/or fission products and an experimental programme of work is now in progress to obtain a comprehensive knowledge of the capabilities of these exchangers. The absorbers selected for this study are:- Titanium Phosphate, Zirconium Phosphate, Manganese Dioxide, Polyantimonic Acid, Hydrous Titanium Oxide and Sodium Copper Hexacyanoferrate.

EXPERIMENTAL STUDIES [2], [3]

The initial assessment showed that a considerable volume of information is available on the use of inorganic ion exchangers for waste stream decontamination. However, the reported data are often of limited use in selecting an absorber for the treatment of a particular waste stream since the candidate material has generally been examined only for a few or individual elements over a small range of operating conditions. Also the information available on α-decontamination is considerably less than that for βγ removal.

Accordingly the main purpose of the current experimental work is to obtain a comprehensive knowledge of the capabilities of inorganic ion exchangers. The performance of the selected absorbers, listed above, is being examined over a range of process parameters including,

- solution pH
- temperature
- contact time
- influence of competing ions, both radioactive and stable, on absorption of a particular element
- effect of γ-irradiation (to 10^9 Rads)

In addition the work is also concerned with

- optimisation of the method of preparation of the absorber to provide material of high capacity and in a form suitable for plant use,
- examination of methods for regenerating spent absorber by treatment with reagents which are themselves acceptable to a disposal process such as vitrification,

- examination of methods for immobilising spent absorber by conversion into a solid form suitable for storage or disposal.

(i) Absorption studies

Table 1 summarises the absorption data obtained so far in experiments conducted at 25°C. In this table the heavy type shows absorbers which have provided decontamination factors (DFs) in excess of 100 for the radionuclides and hydrogen ion concentration shown; and the light type refers to absorbers which have provided DFs between 10 and 100. Where no absorption occurred with any of the selected absorbers this in indicated by the letter X; blank spaces show where measurements have not yet been completed. In all this work, solutions of pH > 1 were 1M in sodium nitrate.

The results obtained so far in the work are showing that good decontamination from fission products can be obtained if a particular absorber is used under certain operational conditions. The results obtained using spiked simulates of waste streams have been confirmed using a solution of irradiated fuel in nitric acid.

Figure 1 illustrates the extent of α-activity removal by different absorbers from solutions of irradiated fuel dissolved in nitric acid and then adjusted to various pH values. It is apparent from the figure that antimonic acid, hydrous titanium oxide and manganese dioxide have a strong affinity for α-emitters over a fairly wide range of pH values.

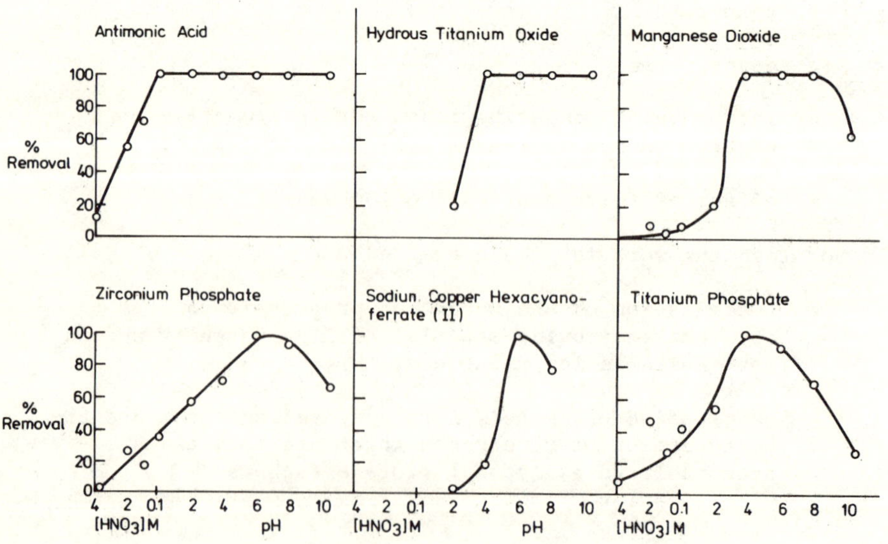

FIG. 1. TOTAL α REMOVAL.

TABLE 1 Summary of Absorber Performance

	4MH$^+$	2MH$^+$	1MH$^+$	pH=1	pH=2	pH=3	pH=6	pH=8	pH=10
Pu	X	X	SbA	**SbA** **HTiO**	**SbA** **HTiO**				
Am	X	X	SbA	SbA		**HTiO** **SbA** **TiP** **MnO**	**SbA** **HTiO**		**HTiO** SbA
Np	X	X	X	X	HTiO				
Cs	TiP	TiP	TiP ZrP	**TiP** ZrP SbA	**CFC** TiP ZrP	**CFC** TiP ZrP	**CFC** ZrP TiP	**CFC**	
Sr			**SbA**	**SbA**	**SbA** **TiP**	**SbA** **TiP**	**SbA** **TiP**	**SbA** **TiP** **MnO** HTiO	**SbA** **HTiO** **TiP** **MnO**
Ru	X	X	MnO	MnO	**MnO** **HTiO**	HTiO MnO CFC	MnO HTiO CFC	**MnO** HTiO	HTiO
Ce	X	X	SbA	**SbA** MnO	SbA MnO	**SbA** **TiP** HTiO	**HTiO** **ZrP** **CFC** TiP	**SbA** **MnO** **HTiO** ZrP	**HTiO** SbA

Legend

SbA = Polyantimonic Acid
HTiO = Hydrous Titanium Oxide
TiP = Titanium Phosphate
ZrP = Zirconium Phosphate
CFC = Copper Ferrocyanide
MnO = Manganese Dioxide

Key

Heavy type: DFs > 100
Light type: DFs 10-100
X: no absorption
Blank: measurements not yet completed.

The absorption of actinides from spiked simulates has been found to be very good for some of the absorbers tested. In general the value of the distribution coefficient is greater for Pu(IV) than for Am(III) and is lowest for Np(V). The presence in the waste solution of di- and tri-valent metal ions such as Fe^{3+}, Al^{3+} and Mg^{2+} reduces the value of the partition coefficient because of competition for exchange sites.

Tables 2, 3 and 4 list the distribution coefficients determined for plutonium, americium and neptunium respectively.

The studies have now been extended to measurements at 40 and 60°C and the results, which are still being assessed, show a wide variation in the performance of individual exchangers and for individual radionuclides.

TABLE 2 K_D values for Pu(IV)

Acidity/ pH	ABSORBERS				
	CSbA	ZrP	MnO	TiP	HTiO
4M	4.660×10^{1}	5.720×10^{0}	1.610×10^{0}	2.740×10^{2}	n.m.
2M	1.220×10^{3}	3.830×10^{1}	1.000×10^{0}	7.720×10^{2}	n.m.
1M	1.340×10^{4}	2.160×10^{2}	3.100×10^{0}	2.890×10^{3}	n.m.
pH 1	5.160×10^{4}	1.150×10^{4}	1.300×10^{1}	8.970×10^{2}	1.020×10^{3}
pH 2	4.740×10^{4}	3.420×10^{3}	1.400×10^{2}	5.620×10^{4}	1.160×10^{4}
pH 7	5.130×10^{2}	n.m.	n.m.	n.m.	n.m.
pH 10	2.370×10^{3}	n.m.	n.m.	n.m.	n.m.
pH 12	7.890×10^{3}	n.m.	n.m.	n.m.	n.m.

All solutions with pH > 1.0 contained 1.0 M $NaNO_3$
n.m. = not measured
25°C

TABLE 3 K_D values for Am(III)

Acidity/ pH	ABSORBERS				
	Polyan S	ZrP	MnO	TiP	HTiO
4M	2.320×10^1	8.930×10^0	1.180×10^1	1.010×10^1	n.m.
2M	1.370×10^2	2.170×10^1	1.150×10^1	1.780×10^1	n.m.
1M	2.800×10^3	2.710×10^1	7.900×10^0	1.060×10^1	n.m.
pH 1	5.590×10^3	2.360×10^1	4.400×10^1	7.900×10^1	2.150×10^1
pH 3	2.720×10^5	1.910×10^3	6.130×10^4	2.230×10^4	1.470×10^5
pH 8.8	9.280×10^4	n.m.	n.m.	n.m.	n.m.
pH 10	1.430×10^5	n.m.	n.m.	n.m.	n.m.
pH 12	1.330×10^4	n.m.	n.m.	n.m.	n.m.

All solutions with pH ≥ 1.0 contained 1.0 M $NaNO_3$
n.m. = not measured
25°C

TABLE 4 K_D values for Np(V) as NpO_2^+

Acidity/ pH	ABSORBERS				
	CSbA	ZrP	MnO	TiP	HTiO
4 M	4.380×10^1	1.530×10^1	1.110×10^1	7.060×10^0	n.m.
2 M	1.980×10^1	1.030×10^1	6.850×10^0	4.700×10^0	n.m.
1 M	2.920×10^1	1.170×10^1	7.800×10^1	~ 0	n.m.
pH 1	1.540×10^2	8.450×10^0	1.130×10^1	5.470×10^1	5.940×10^0
pH 2	1.800×10^2	5.065×10^1	1.740×10^1	5.310×10^2	1.980×10^3

All solutions with pH ≥ 1.0 contained 1.0 M $NaNO_3$
n.m. = not measured
25°C

(ii) Radiation stability

The six inorganic ion exchangers under investigation have been subjected to γ-irradiation in the Harwell spent fuel cooling pond. The samples were irradiated at 1.44 Mrads per hour to total γ-doses of 28, 61, 129, 219, 500 and 1000 Mrads; the temperature of the samples during irradiation was 37°C.

Polyantimonic acid, titanium phosphate, sodium copper hexacyanoferrate(II), manganese dioxide and zirconium phosphate were irradiated dry whereas hydrous titanium oxide was irradiated in water.

After irradiation the samples were weighed and then conditioned to a given pH value before being contacted with a simulated waste solution containing fission products and actinides.

The results obtained in the batch contact experiments show that γ-irradiation to the total doses mentioned has little or no effect on the efficiency of these absorbers for the removal of radionuclides from solution.

(iii) Absorber preparation

Titanium phosphate and polyantimonic acid have been prepared by various methods and have been characterised by thermal analysis and X-ray diffraction methods. The ion exchange performance of the different preparations is being tested.

Several methods of preparing manganese dioxide absorber are being examined and the performance of the products compared with that of a commercially available material (MANOX - Applied Research, Belgium). The best material obtained so far was prepared by treating a manganous sulphate solution with ammonium hydroxide. The exchange kinetics of this material for βγ emitting nuclides were similar to those of the commercial product but the decontamination factors were rather better. Work to optimise preparative conditions is continuing.

A simple heat treatment during the preparation of sodium copper hexacyanoferrate(II) has been found to improve significantly the chemical stability of the product. Untreated material decomposes in solutions of pH $>$ 10, producing sodium ferrocyanide and copper hydroxide. If the freshly precipitated cyanoferrate is heated at ~ 80°C for two hours the product has been found to be stable in solutions with pH values up to 12.

ACKNOWLEDGEMENTS

The experimental data presented in this paper was obtained by B.A. Phillips, S.P. Dagnall and N.P. Monckton of Separation Processes Group, Chemical Technology Division, AERE, Harwell. Their contribution is gratefully acknowledged.

This work was commissioned by the UK Department of the Environment as part of its radioactive waste management research programme. The results will be used in the formulation of Government policy but at this stage they do not necessarily represent Government policy.

REFERENCES

[1] E.W. Hooper, B.A. Phillips, S.P. Dagnall, N.P. Monckton. "An assessment of the application of inorganic ion exchangers to the treatment of intermediate level wastes", AERE R11088, May 1984.

[2] E.W. Hooper. "The application of inorganic ion exchangers to the treatment of alpha-bearing waste streams", Paper to IAEA Technical Committee Meeting on Inorganic Ion Exchangers and Adsorbents for Chemical Processing in the Nuclear Fuel Cycle, Vienna, June 1984, IAEA TC-518-10.

[3] B.A. Phillips, E.W. Hooper, S.P. Dagnall, N.P. Monckton. "Study of the behaviour of inorganic ion exchangers in the treatment of medium active effluents. Part 1: Preliminary experimental work on absorber preparation and performance", AERE G2872, November 1984.

The Use of Coordinating Copolymers to Extract ^{106}Ru from Simulated Nuclear Wastes

Michael J. Hudson, Department of Chemistry, University of Reading, Whiteknights, P.O. Box 224, Reading, RG6 2AD
Alan Dyer, Department of Chemistry, University of Salford, Salford M5 4WT

SUMMARY

The extraction of ^{106}Ru from simulated waste and pond waters has been investigated in a preliminary manner. A dithiocarbamate copolymer has been found to be of potential use for pond water (K_D 13,800). A thiadiazole copolymer (K_D 1550) and a thiol copolymer (K_D 800) are of potential use for waste waters. Low molecular weight precipitating agents with two separate donor atoms have some use for acidic waste waters (e.g. 1,3,4-thiadiazole-2,5-dithiol) (K_D 800) and for the basic pond water (e.g. amidinothiourea, K_D 2460). These substances appear to be stable to radiolytic decay after extracting ^{106}Ru from a 3.6×10^{-6} molar solution for periods up to four weeks. There are indications that in dithiocarbamate copolymers the metal (Ru) requires six donor sulphur atoms (as is the case with Rh). It is suggested that copolymers which are cross-linked with divinylbenzene are too inflexible to allow the metal to have six-fold coordination. Thiols may not have this problem.

INTRODUCTION

This study is one in a series [1-11] in which the interactions of precious and base metals with coordinating copolymers are being studied. Coordinating copolymers are those compounds which are able to form a bond, which has a significant amount of covalent character, with metals. It is possible to classify coordinating copolymers according to the scheme in Figure 1. The ion-exchangers have been separated from the coordinating copolymers but there is often a dual behaviour. The coordinating copolymers have covalently

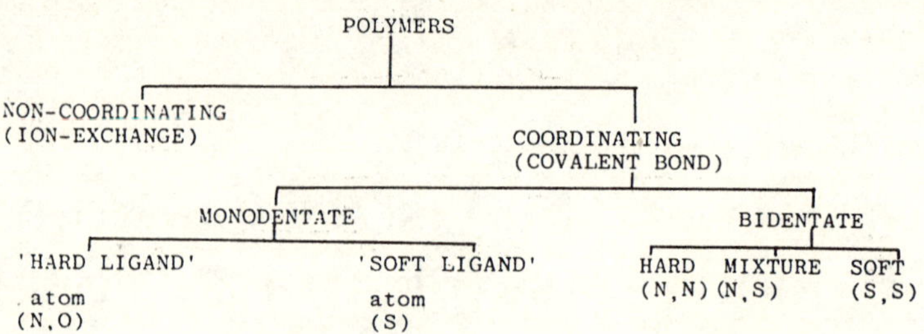

FIGURE 1

bound ligand groups which have one or more donor atoms. These groups may be generally classified into monodentate, N (e.g. pyridine, amines), O (phenols, ethers); S (thiols) or bidentate N,N (amines); N,S (thiols/amines); S,S (xanthates, dithiocarboxylates). There are twenty commercial coordinating copolymers and a further large number of experimental copolymers from which to make a selection. With respect to the treatment of nuclear wastes, coordinating copolymers have potential in a number of areas which include.

TREATMENT OF POND WATER.

Treatment of liquid effluent wastes.

Coordinating copolymers should be regarded as versatile reagents. They can be made selective for particular types of metals such as actinide separation from lanthanides; precious metals from base metals or precious metals from actinides. The copolymers can be soluble in water (high concentration of hydrophilic groups) or insoluble (e.g. based on poly(styrene). They can be used as flocculants, or as insoluble beads. Previous work with non-radioactive transition metals has indicated that coordinating copolymers may be used to separate Fe/Co [3]; Ag from base metals [6]; rhodium from base metals [5]; cadmium from zinc [8]; rhodium from iridium [2]. Metals can be extracted in strong acids as well as nearly neutral solutions but in the latter use the coordinative mechanism is more likely to dominate. Scatchard plots, using data from ion-selective electrodes, have been used to determine the formation constants and the number of sites to which cadmium may bond [8].

The transition metals Ru, Rh, Pd, Tc and Mo are present in irradiated fuels in the form of a single phase alloy [12-14].

The fission yields of these metals are quite significant. For example the elemental fast fission yields for ^{239}Pu (365 days irradiation and 180 days cooling time) are quoted as Ru 23.1%; Rh 7.0%; Pd 15.0%; Tc 6.0%; Mo 22.9%. Each of these elements has a well established coordination chemistry. Since the metals have an affinity for sulphur donor atoms, particularly when the metal is in a low oxidation state, it was decided to study the extraction of ^{106}Ru using coordinating copolymers [15]. The normal oxidation state for ruthenium is + 3 and cationic species are well known [16]. After the treatment of the nuclear fuels with nitric acid the ruthenium usually retains the nitrosyl group (Ru-N-O) [17]. Consequently, the coordinating copolymers need to combine with these nitrosyl complexes. Although the nitrosyl group itself is unreactive it does not appear to inhibit substitution in other parts of the octahedral complex. The position opposite to the NO group is labile. Dithiocarbamate complexes $RuNOL_3$ ($L=Et_2NCS_2^-$) have been prepared [18] and it was for this reason that copolymer ((8)) was studied. This copolymer (PIED) has been used to separate Fe/Co [3]; Ag [6]; rhodium from base metals [5] and cadmium from zinc [8]. The previous studies of the extraction of rhodium with PIED is pertinent to this paper because the chemistry of ruthenium(III) and rhodium(III) are rather similar. It has been shown that the rate of extraction of rhodium is quite fast ($t_{\frac{1}{2}}$ is under five minutes). It should be appreciated that PIED is an uncrosslinked copolymer in a powder form and that the copolymer contains approximately 25% by mass of sulphur. The donor sulphur atoms are largely at the ends

FIGURE 2 Rate of Extraction–Rhodium as a Function of Particle Size
A. 150-250 B. 250-500 C. Over 500m.10^{-6} (mesh)

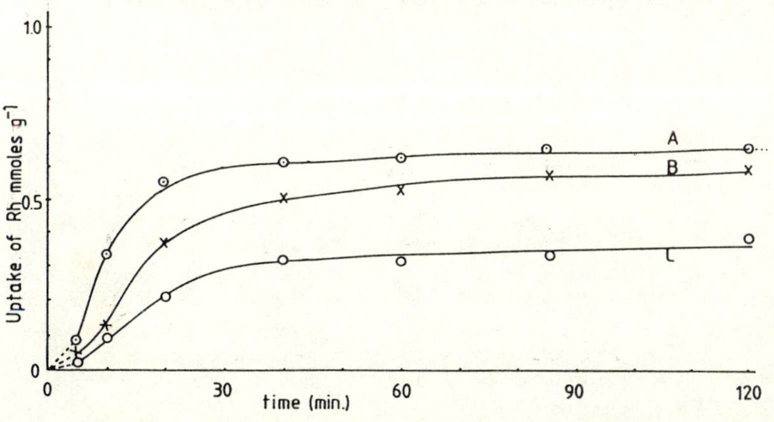

of the branches. The rate of extraction of rhodium from chloride media is illustrated in Figure 2. Clearly the rate of extraction and the final capacity are a function of the size of the particles. The extraction was done by equilibrating (PIED 10.2 g) with sodium 'hexa'chlororhodate(III) in hydrochloric acid (4M HCl; Rh 0.01 dm^{-3}, 25 cm^3). Both the rate and capacity of the PIED for rhodium is dependent upon the available surface area of the bead. Since the metal forms covalent bond with PIED, the copolymer is cross-linked near to the surface and further diffus ion of metal ions is severely limited. A general rule appears to be that coordinating copolymers require small particle size for maximum rates and capacities. Although it will be shown later on that ^{106}Ru is not efficiently extracted from acidic waste-waters from nuclear plants this is only true when PIED itself is left in acidic solutions unbound to metals. Like all polyelectrolytes PIED is more stable as a metal complex or salt. The ^{106}Ru is at low concentration so that PIED slowly decomposes in acidic solution. This problem does not arise with the basic pond water solutions. Spectral studies [1,2] have shown that in Rh-PIED the Rh atom is surrounded by six sulphur atoms. Whereas PIED has a capacity of ca. 0.5 mmoles Rh $g.^{-1}$, a copolymer cross-linked with divinylbenzene has a low capacity (e.g. IR-45 with CS_2 in sodium hydroxide. Consequently, it is thought that Rh is extracted only when the copolymer can bend sufficiently to enable the six donor-S atoms to bond directly to rhodium such that the Rh-S_6 moiety approximates to an octahedron.

Electron spin resonance studies (e.s.r.) [2] of Cu-PIED compounds have shed further light on the nature of the coordination of metals to the copolymers. The e.s.r. spectra are illustrated in Figure 3. The first point of note is that the spectra are all clear, single signals with

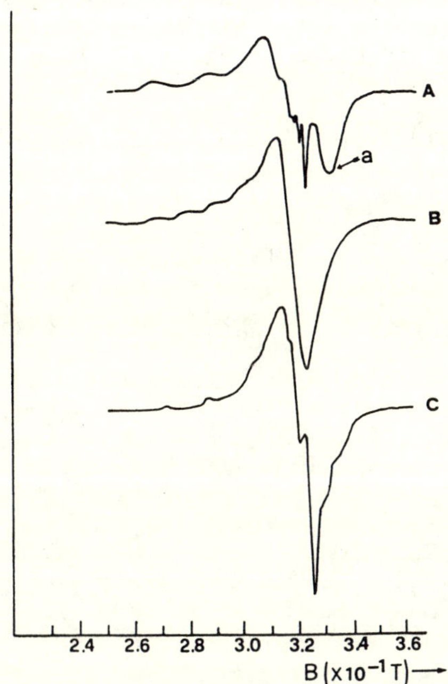

FIGURE 3

X-band e.s.r. spectra of Cu^{II} loaded materials (reference 2)-at 77K.

A: 2-4% Cu^{II}-IR45CS_2H (a denotes the overshoot line which is characteristic of dithiocarbamates.

B. High loaded ca.15% Cu(II)-PIED

C. Low loaded ca.1-2%Cu(II)-PIED

developed fine structure. We are dealing with well defined sites involving S or N, S_n or N coordination and not with adsorption. No half-field lines [19,20,21] were observed and consequently each metal has its own separate coordination sphere. In IR45-CS_2H the Cu has a $g_{//}$ value of 2.126

TABLE 1 Spectra of Cu^{II}-loaded Materials at 77K

Material	$g_{//}$	$g_{\perp}$	$A_{//}$ [a]
A. Cu^{II}-IR45CS_2^- r.t.	2.135(2)	2.025(2)	156(1)
B. Cu^{II}-PIED (high-loaded)	2.236(3)	2.074(3)	104
C. Cu^{II}-PIED (low-loaded)	2.209(3)	2.065(3)	154(2)8

a. x 10^{-4} cm^{-1}

(Figure 3A) which, together with other evidence (e.g. bands at 16,000 and 21,800 cm^{-1}) suggests a distorted CuS_4+S moiety. In high-loaded Cu-PIED (Figure 3,B) $g_{//} > g_{\perp} > 2.02$ and there is a low $A_{//}$ parameter (104 x 10^{-4} cm^{-1}) which suggests a pseudo-tetrahedral coordination CuN_3S or CuN_2S_2. This is a convenient model for copper-blue protein. The low-loaded Cu-PIED has a different geometry - perhaps a trigonal bipyramidal CuN_2S_3.

There is a direct spectroscopic evidence [15] that the coordination of the metals to the copolymers causes strain. The observed spectral lines lie close to those for dialkyl-phosphoro- and dialkylphosphino-dithioates and at significantly lower energies than low molecular weight dithiocarbamates which are known to have shorter bond lengths. The difference in between the corresponding lines approximates to 10^3 cm^{-1} molecule or 1kK molecule $^{-1}$ [23]. This enables a very approximate strain energy to be calculated as 12 J mol^{-1}. More studies need to be done to estimate the maximum permissible strain energy which here refers to the loss of the ligand field stabilisation energy when going from the low molecular weight compound to a 'corresponding' high molecular weight

(M-copolymer) compound.

RESULTS AND DISCUSSION

Three simulated solutions were studied [15]. The first two solutions have a mixture of nitric acid, sodium nitrate

Table 2 Composition of simulated waste solutions.

	Solution A	Solution B	Solution C
NaOH	—	—	0.005 M
Na_2CO_3	—	—	0.0028 M
HNO_3	0.15 M	0.08 M	—
$NaNO_3$	5 M	2.5 M	0.01 M
Ru-NO[a]	3.63×10^{-6} M	3.63×10^{-6} M	3.63×10^{-6} M

[a] Ru-NO signifies species which contain this group.

and the ruthenium-nitrosyl compound. Solution A is more acidic than B and has a higher ionic strength. The ^{106}Ru concentrations used were higher than those in real effluent but were fixed for the purposes of this preliminary study for convenience of analysis. Solution C approximates to 'pond water' and is a basic solution. The distribution coefficients (K_D) were calculated for solid substances which included non-polymeric materials, commercial copolymers and new copolymers. The solids (0.2 g) were contacted with the solutions (20 cm^3) for one week in poly(ethene) vials with constant stirring. Control solutions with no equilibrated solid phases were treated in the same way. ^{106}Ru was determined by Cerenkov counting via its Rh daughter. The K_D values were calculated [13] from equation (1), where A_O = initial activity of solution, A_F = final activity, V = solution (cm^3), W = mass of resin (g).

$$K_D = \frac{A_O - A_F}{A_F} \frac{V}{W} \quad (1)$$

The poly(vinylpyrrolidone) material decomposed to a gel. The chemical formulae of the simplified repeating unit of ((8)) is given in Figure 4. Substance ((1)) is insoluble in solution A but is partially soluble in solutions B and particularly C (Table 2). Substance (7) (amidinothiourea) is unstable in acid solutions (A,B). The dithiocarbamate copolymer (PIED) substance 8 is also unstable in these solutions.

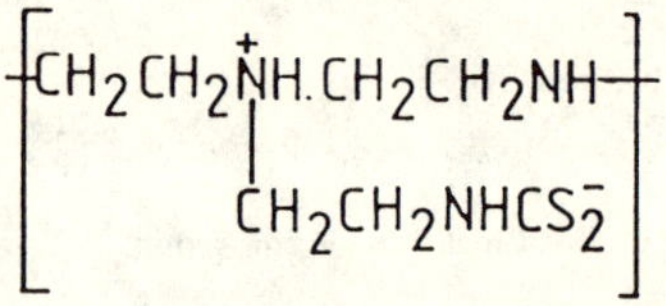

$$\left[CH_2CH_2\overset{+}{N}H.CH_2CH_2NH \atop \quad\quad | \atop \quad\quad CH_2CH_2NHCS_2^- \right]$$

The Repeating Unit of PIED

FIGURE 4

The copolymer ((9)) was made by reaction of chloromethylated poly(styrene) [6,19] with substance (1). The advantageous cross-linking in this copolymer will be mentioned shortly. The results for the extraction of ^{106}Ru are given in Table 3. With respect to substance ((1)) the K_D values for solutions B and C are certainly greater than unity which indicates some capacity for ^{106}Ru. The substance ((1)) does not appear to be able to precipitate ^{106}Ru from these two solutions in which it has some solubility. The K_D value in solution A reflects the insolubility of this substance in the stronger acidic medium and perhaps the ability of ruthenium to cross-link the two thiol groups to produce an insoluble copolymer. It is particularly interesting that the copolymer ((9)) derived from ((1)) has higher K_D value for each of the solutions. This particular copolymer has 2-4 per cent cross-linking and is prepared from XE-305 which is an experimental resin from Rohm and Haas. The beads have a character which is intermediate between gel and macroreticular (macroporous). The resin has long alkyl chains which cross-link polymer particles. The particular advantage of the resin is that it expands sufficiently to allow groups to be

TABLE 3 Extraction of ^{106}Ru with solid reagents. Apparent K_D values ($cm^3\ g^{-1}$).

Sample (physical form)	Solution A	Solution B	Solution C
(**1**) 1,3,4-Thiadiazole-2,5-dithiol (powder)	800	80	10
(**2**) An aminophosphonic acid copolymer (beads)	100	100	36
(**3**) A thiol copolymer (beads)	800	530	70
(**4**) An iminodiacetate copolymer (beads)	480	300	1.1
(**5**) An amideoxime copolymer (beads)	70	95	103
(**6**) A pyrrolidone copolymer (powder)	65	1.5	—
(**7**) Amidinothiourea (powder)	200	512	2 460
(**8**) A dithiocarbamate copolymer powder	250	250	13 800
(**9**) A thiadiazole copolymer (beads)	1550	260	16

bonded to the matrix and still leaves sufficient room and flexibility in the copolymer for diffusion and bonding of the ruthenium. It is likely that ion-exchange and coordinating mechanisms both operate. It is not clear at this stage why the copolymer ((9)) is more efficient in acidic solutions than in basic ones. Since the results for the three solutions are in the same general order as the thiol copolymer ((3)) it appears likely for ((3)) and ((9)) that the sulphur atom is bound to the ruthenium atom. The difference in solution A may be because there is S, N coordination between compound ((9)) and ruthenium but just S bonded in ((3)). However, in separate studies we have noted that copolymers such as ((3)) with significant amounts (ca 6%) of cross-linking by divinylbenzene, are too rigid to enable the copolymer chains to move in order that the metal may achieve the required stereochemistry (e.g. octahedral) and the correct number of coordinated atoms. If the strain energy on the metal by the copolymer is significantly above the approximate value of ca 10J $mole^{-1}$ which was mentioned earlier for Cu-PIED (Cu-copolymer ((8))) then the copolymer may reject the metal. In nature metals are bound to copolymers which are cross-linked by hydrogen-bonds which are much weaker than the above covalent bonds. In addition, the aromatic ring in divinylbenzene contributes to the rigidity or lack of flexibility of the copolymer.

The compound, amidinothiourea ((7)), and the dithiocarbamate copolymer PIED ((8)) appear to be efficient at removing

^{106}Ru from pond water (C) but the mechanisms by which this is done are probably different. The low molecular weight amidinothiourea forms a complex with ^{106}Ru which then probably acts as a bridging agent between ligands. The insoluble, cross-linked product appears to be stable in basic media. Previous work with PIED ((8)) has suggested that it has a good capacity for precious metals (Ag(I) capacity is 1 g of metal for 1 g of dry copolymer) but that it is unselective. In the case of ^{106}Ru the copolymer has a high K_D value (13,800) and this implies that the copolymer could be used in pond water to extract this and other metals which form dithiocarbamate compounds. The dithiocarbamate group is one of the strongest ligands and it has even been possible to stabilise Ni(IV) on the copolymer by oxidation of Ni(II)-PIED with bromine [3]. Therefore, it is likely that the ^{106}Ru is bound to the copolymer through the donor sulphur atoms. This is the case with rhodium(III) which is chemically similar to Ru(III) [16]. The nitrosyl group may be retained as in the low molecular weight compound [18]. Although the ^{106}Ru-PIED compound appears to be stable to radiolytic decay and to chemical attack by acids further studies are in progress. The significantly lower K_D value (250) in solutions A and B is considered to be due to decomposition of PIED prior to combination with ruthenium. It might, in future, be possible to use Ni-PIED in which the metal is bound to four sulphurs. However, in Cu-PIED, there is pseudo-tetrahedral coordination and consequently ruthenium and copper compete for different sites. This is to say that PIED may be selective for Rh over Ni or Ru over Ni but it is not likely to be selective for Ru over Cu. Previous work, [1,2] showed that PIED is unselective in Rh/Cu solutions.

Clearly the copolymers are selective for ^{106}Ru over sodium ions even when there are high (5 molar) concentrations of sodium nitrate. It is of interest that four of the materials which include copolymers ((3)) (thiol); ((4))(iminodiacetate); ((9)) thiadizole; appear to have higher capacities in solution A than in solution B, even though solution A has a higher sodium ion concentration.

CONCLUSION

If coordinating resins are to be used to extract radionucleides such as ^{106}Ru, particular attention needs to be paid in future to:-

1. Susceptibility to radiolytic damage [22] from alpha and beta particle emitters.

2. Rates of absorption - as coordinating copolymers may react slowly.

3. Possibility of continuous plant operation.

4. Possibility of specialised applications - where there are low levels of emission.

These and other aspects are currently being studied in a joint project between the two authors of this paper and British Nuclear Fuels plc.

REFERENCES

1. Ellis A.F., Ph.D. Thesis, Reading, 1984.

2. Ellis A.F., Hudson M.J., Tomlinson A.A.G., Dalton 1985 accepted for publication.

3. Giwa C.O., and Hudson M.J., Hydromet., 1982, 8, 65.

4. Hudson M.J., Hydromet., 1982, 9, 149-168.

5. Hudson M.J., and Thorns J.F., Hydromet., 1983, 11, 289.

6. Hudson M.J., and Shepherd M.J., Hydromet., 1983, 9, 223.

7. Hudson M.J., Dyer A., et al. Chem. Comm. 1984, 21, 1457-8.

8. Hudson M.J., and Tiravanti G., Die Makromol. Chem., 1985 (in press).

9. Hudson M.J., Dalton, 1985, paper accepted.

10. Hudson M.J., and Glaves L.R., Dalton, 1985, paper accepted.

11. Hudson M.J., and Shepherd M.J., Hydromet., 1985 (paper in press).

12. Horspool J.M., Parkinson N., Findlay J.R., Potter P.E., Rand M.H., Russell L.E., Batey W., Fuels and Fuel Elements for Fast Reactors, Proceedings of International Symposium, Brussels, 1973 (IAEA, Vienna, 1974), I, 3.

13. Huber H., Kleykamp H., GfK report KFK, 1324, 1972.

14. O'Boyle D.R., Brown F.L., Sanecke J.E., J. Nucl. Mats., 1969, 29, 200.

15. Dyer A., Keir D., Hudson M.J., and Leung B.K.O., J. Chem. Soc., Chem. Commun., 1984, 1457.

16. Greenwood N.N., and Earnshaw A., Chemistry of the Elements, Pergamon, 1984.

17. Boswell G.G.J., and Soentono S., J. Inorg. Nucl. Chem., 1981, 43, 1625.

18. Domenicano A., Vaciago A., Zambonelli L., Loader P.L., and Venanzi L.M., Chem. Commun., 1966, 476.

19. Leung B.K.O., Ph.D. Thesis, University of Reading, 1985.

20. Hathaway B.J., and Billing D.E., Coord. Chem. Rev., 1970, 5, 143.

21. Hathaway B.J., and A.A.G. Tomlinson, Coordin. Chem. Rev., 1970, 5, 1.

22. Hathaway B.J., Coordin. Chem. Rev., 1982, 41, 432.

23. Simon G.P., Stability of Ion-Exchangers in Ionising Radiation, in Ion-Exchange for Pollution Control, Editors Calmon C., Gold H., and Prober R., CRC Press, Florida, 1979.

24. Tomlinson A.A.G., J. Chem. Soc., A, 1971, 1409.

ION EXCHANGE IN FUEL POND AND WATER TREATMENT

Ion Exchange in Candu Reactor Systems

B.R. Nott and S.R. Goodwin, Ontario Hydro, Toronto, Canada

An overview of the role of ion exchange in the various purification circuits of CANDU-PHW* reactors of Ontario Hydro is presented. Ion exchange systems used for purification of the heavy water in the primary heat transport (PHT) and moderator systems and for purification of the irradiated fuel bay water are described in detail. Information regarding water chemistry and other operating conditions in the various circuits is also presented along with some operating experience. The facilities and efforts of Ontario Hydro's Research Division for supporting the operation of the ion exchange systems of Hydro's CANDU reactors are described. Short and long term work programs for improving performance of the ion exchange systems are outlined.

INTRODUCTION

CANDU-PHW reactors use natural uranium fuel and heavy water (D_2O) for the moderator and coolant. There are two separate circuits; the low pressure, low temperature moderator circuit and the high pressure and high temperature primary heat transport (PHT) circuit. This separation simplifies specification of optimum chemistry conditions in each of the systems. Purification of waters in the various circuits is necessary to enable satisfactory reactor

---

* CANada Deuterium Uranium-Pressurized Heavy Water

operation. Ion exchange is the primary unit operation used in the purification circuits. Ion exchange resins also play a significant role in chemistry control of the various aqueous systems. The primary objectives of reactor circuit chemistry control are to minimize corrosion, maintain low radiation fields and control reactivity.

Reviews of water chemistry of CANDU-PHW reactors have previously been published[1,2]. In this paper, attention is focussed on the ion exchange systems used for purification of the various aqueous circuits of CANDU reactors. The systems used for purification of heavy water in the PHT and moderator circuits and of the natural (light) water in the irradiated fuel bay (IFB) are described in detail. These systems account for a majority of the total ion exchange resin usage at a station. Nuclear grade ion exchange resins are used in the purification systems. Some of the other systems using ion exchange (but not discussed here) include heavy water collection systems, the main boiler feedwater treatment plant and the reactor decontamination systems.

Much of the information presented here applies to all of the CANDU nuclear stations of Ontario Hydro - Pickering NGS 'A' and 'B', Bruce NGS 'A' and 'B' and Darlington NGS. Each of these is a four unit station.

PURIFICATION SYSTEMS

PHT Purification System

A schematic diagram of the PHT system and its purification circuit is shown in Figure 1. Heavy water is circulated in the main system at approximately 10 MPa and 310°C. The system materials are carbon steel piping, zirconium alloy fuel cladding and pressure tubes, and nickel alloy steam generator tubing.

The PHT system becomes contaminated with tramp fission products (radioiodines, radiocesium, etc), corrosion products, and trace impurities from the make-up heavy water. A small portion of corrosion products from system surfaces is transferred to the reactor core where some of them are activated by the neutron flux and are carried back to out-core areas where they are incorporated in the oxide layer on system surfaces. As a result of this 'activity transport' phenomenon, radiation fields are produced around the PHT system. The activated corrosion products include Co-60, Cr-51, Zn-65, Mn-54 and Zr-95.

The PHT purification system consists of filters and ion exchange columns located in a bleed purification circuit taken from the main PHT flow as shown in Figure 1. The flow through the bleed circuit is cooled to about 40°C before it enters the purification columns. The function of the purification system is to remove fission products as well as the soluble and insoluble corrosion products in order to help control activity transport and minimize build up of out-reactor radiation fields. The ion exchange columns are also used to maintain the pH* of the PHT water in the alkaline range of 10.3 - 10.7 to minimize corrosion. Removal of ionic impurities such as chlorides and carbonates is also controlled by the ion exchange columns. Chlorides can induce stress corrosion cracking of stainless steel components and the ion exchange resins themselves are thought to be the primary source of chlorides; hence low chloride ion exchange resins are specified. Carbonates originate from the radiolytic decomposition of organic materials such as oil and resin fines and from air-ingress during some shutdowns.

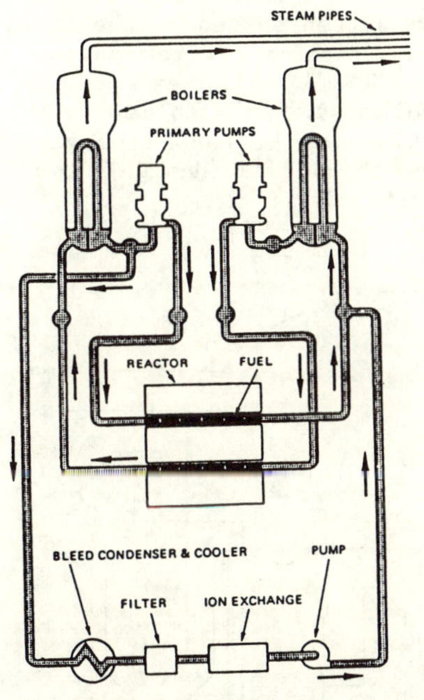

FIGURE 1 PHT Purification System

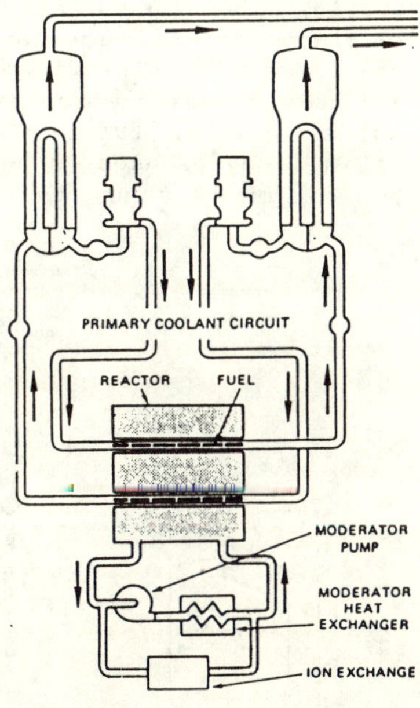

FIGURE 2
Moderator Purification System

* In D_2O systems, the 'apparent pH' is measured using standard pH instruments calibrated in H_2O solutions. In this paper, pH refers to 'apparent pH' for D_2O systems.

A small excess of dissolved deuterium (3–7 cc/kg D_2O) is maintained in the PHT water to scavenge oxygen to minimize corrosion of zirconium alloy and carbon steel components.

Mixed bed resins, in Li^+/OD^- form (1:1 equivalent mixture - Amberlite, IRN-154 LC) are used for normal PHT purification and for maintenance of alkalinity. A D^+/OD^- form mixed bed resin (Amberlite, IRN-150 LC) is used for Li^+ removal from the PHT water should the pH exceed specifications (usually a start-up phenomenon). The resins are deuterated prior to use, in a hopper with a slow upflow of D_2O. (This is to prevent downgrading of the D_2O in the system.) The deuterated resin is usually transferred to the service ion exchange column in a closed loop slurry operation using D_2O, as shown in Figure 3.

Each ion exchange column is a cylindrical vessel (2 ft. diameter) with a conical bottom, and has two well screens, one at the top near the centre of the resin bed and the other located centrally in the bottom of the conical portion. See Figure 3. The bottom screen serves as the resin retaining screen. The conical bottom facilitates removal of spent resin by slurrying (Pickering NGS 'A' is an exception and uses disposable ion exchange columns). Each column holds 7 cu. ft. (198 L) of resin. Up to eight ion exchange columns may be used in parallel and the normal (design) flow per column is about 2.7 L/s (or 48 BV/h).

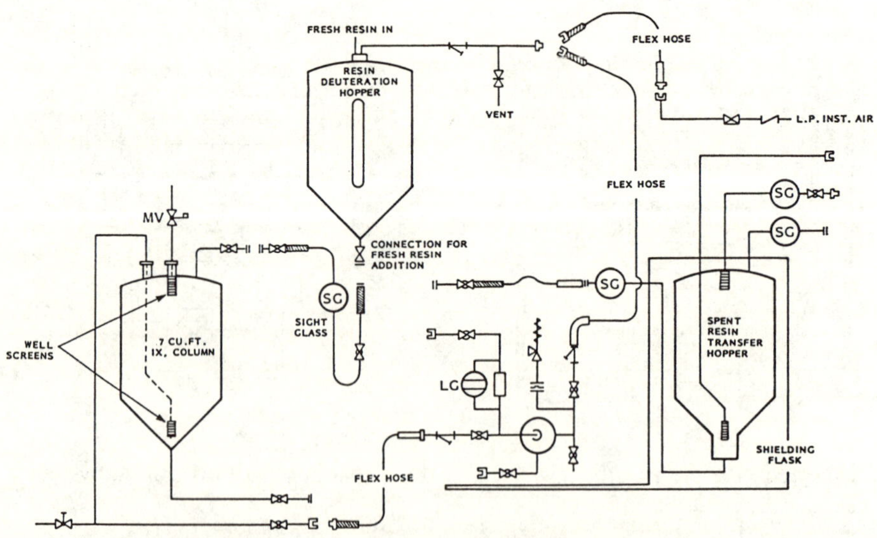

FIGURE 3 Slurry Transfer of Ion Exchange Resin

Fission product iodines (I-131, I-133) are kept under control, between 3-5 μCi/kg D_2O. Fuel integrity in Hydro's CANDU reactors has been exceptionally good. Chloride concentration in the PHT is kept below 0.1 mg/kg D_2O.

The Li^+/OD^- mixed bed column is declared 'spent' if chloride concentration in the column effluent exceeds 0.1 mg/kg D_2O or if I-131 in the column effluent exceeds one-tenth of that in the influent. The spent PHT resin is slurried out of the column, de-deuterated and then transferred to in-station storage tanks.

Moderator Purification System

A schematic diagram of the moderator purification system is shown in Figure 2. Heavy water is circulated in the moderator system at 70°C. The major moderator components are stainless steel piping and vessels, zirconium alloy calandria tubes and nickel alloy heat exchanger tubes.

The function of the moderator purification system is to purify the moderator D_2O to minimize concentration of ionic impurities. Some ionic impurities enhance net radiolytic decomposition of D_2O (into D_2 and O_2 gases) and increase the risk of D_2 excursions in the moderator cover gas (helium) and of corrosion from the oxygen. The moderator purification system is also used to control the reactivity worth of the moderator by removing soluble neutron poisons (B and Gd) in a controlled manner.

Boron, as boric acid, is used in the moderator for long term 'shim' control to counteract excess reactivity in new fuel. As the fuel reaches equilibrium conditions, boron must be removed from the system. Because the capacity for B uptake by ion exchange resin is dependent on B concentration in solution, economical use of resin to fully utilize its boron capacity requires a multi-column approach, with each column operating in its own range of B concentration (say >0.5 mg B/kg D_2O for one column and $\leq$0.5 mg B/kg D_2O for the second, in a two column approach).

Gadolinium is used (at all stations except Pickering NGS 'A') for two purposes. It is added as a solution of $Gd(NO_3)_3 \cdot 6H_2O$ in heavy water. A small concentration (approx 1 mg Gd/kg D_2O) is used to compensate for the absence of equilibrium levels of Xe-135 in the fuel on reactor start-up after a shutdown. (B is used for this purpose at Pickering NGS 'A'.) Gd is chosen for the purpose because it burns out at almost the same rate as Xe-135 builds up after reactor re-start thus requiring little adjustment of Gd concentration. The Second Shutdown System (SDS2) also uses

Gd. When this system is activated, Gd is added to the moderator to shut the reactor down. Gd concentration in the moderator under these conditions is about 15 mg Gd/kg D_2O. After activation of this system, it is necessary to remove the Gd from the moderator to enable a reactor re-start. Gd precipitates at a pH just above 7.0; therefore it is important to avoid contamination of the moderator with alkali.

Moderator purification is accomplished by diverting a small portion of the main moderator circuit flow through the ion exchange purification circuit. Strong acid cation/strong base anion mixed bed resins in D^+/OD^- form (Amberlite, IRN-150) are used for boron adjustment as well as for purification. Strong acid cation/weak base anion mixed beds (eg, Amberlite, IRN-300) are used for Gd removal and for normal moderator purification. B is not removed by this mixed bed resin because of its weak base anion component. The resins are deuterated prior to use. Up to six ion exchange columns may be used in parallel. Each column holds a 7 cu. ft. (198 L) resin charge and is similar in design to the PHT column described earlier.

Normal moderator purification involves removal of ionic impurities such as nitrates and various corrosion products. Moderator conductivity is kept below 0.3 mS/m and the resin is declared 'spent' if the column effluent conductivity exceeds 0.1 mS/m or if the chloride concentration in the effluent exceeds 0.1 mg Cl^-/kg D_2O. Spent moderator resins are dedeuterated prior to their discharge to spent resin storage tanks.

The performance of the moderator purification system of Ontario Hydro's CANDU reactors has been quite good. However, some problems have been experienced at Bruce NGS 'A' during resin slurry transfer operations and from plugging of column well screens and in-line strainers. Problems were also experienced from unsatisfactory operation of ion exchange resins used for Gd removal. These problems were solved as described later in this paper.

IFB Purification System

The irradiated fuel discharged from CANDU nuclear reactors is stored under water in irradiated fuel bays (IFB's). The water in the bays provides cooling to remove decay heat and provides shielding from the radioactivity of the IF bundles. Water purity in the IFB's must be controlled to (i) minimize radiation fields and radioiodine levels in the bay area to reduce hazard to operating personnel, and (ii) maintain clarity of the water for ease of operation.

The IFB's in Ontario Hydro stations are designed to provide adequately cooled and shielded storage for the irradiated fuel normally produced by four reactors (over a station's life time) plus a sudden discharge of a complete load of fuel from one reactor. Heat exchangers are provided for removal of the heat load.

The IFB purification system consists of filters, ion exchangers, skimmers and vacuum equipment. Mixed bed ion exchange resins in H^+/OH^- form (Amberlite, IRN-150) are used. Three to five columns are used (depending on the station) for IFB purification. Each column holds a resin charge of 1700 L and the flow per column is about 38 L/s (or 80 BV/h). The ion exchange columns are designed to be manually loaded with fresh resin and spent resin is slurried to in-station storage tanks.

A recent review[3] of the chemical control performance of Ontario Hydro's IFB's revealed that good chemical control has been maintained and that the effect of the bay water contamination on the long term IF cladding integrity was insignificant. The ranges for some of the important parameters were: pH = 5.8 - 9.0, conductivity < 0.2 mS/m, Cl- < 0.3 mg/kg H_2O, I-131 < 20 μCi/kg H_2O.

SPENT ION EXCHANGE RESINS

Spent resins from the purification systems are transferred from in-station storage tanks to 3 m^3 carbon steel cylindrical liners and shipped to the Bruce Nuclear Power Development (BNPD) waste storage site for long term storage. Depending on the radiation field from the resins, the liners are stored in Quadricells (>15 R/h field) or in trenches (≤15 R/h). A detailed account of Ontario Hydro's radioactive waste management practices has previously been published[4]. Volume reduction of spent ion exchange resin wastes is not practised at the present time.

ION EXCHANGE RESEARCH

The Water Research Unit of Ontario Hydro's Research Division offers a whole range of ion exchange services from trouble shooting ion exchange systems to long range research to improve ion exchange system performance. The division has laboratory scale as well as full scale ion exchange test facilities. The full scale facility includes two 8 cu. ft. (226 L) ion exchange columns (similar in design to those used in many of Hydro's nuclear purification systems) and a recirculating test loop with a flow capacity of 10 L/s. There is also a provision for resin slurry transfer between the two columns.

The Unit's ion exchange work programs comprise short and long term projects.

Short Term Programs: These are directed at solving immediate problems. For instance, laboratory tests were carried out to evaluate several candidate resins as substitutes to Amberlite IRN-300 (its production was discontinued in 1983) mixed bed resin used for Gd removal from the moderator; a provisional substitute was identified and has been used at the stations.

Long Term Programs: These focus on improvements in the various ion exchange systems and equipment realizable over a number of years. Some examples follow:

1. Resin Utilization: It was established[5] from tests in our full scale ion exchange facility, that resin utilization is poor near the walls of the cylindrical columns in the lower portions of the resin beds, especially at high flow rates. This was attributed to the particular column design (conical bottom and a small centrally located outlet well screen. See Figure 3). Further work will focus on means of improving resin utilization.

2. Resin Slurry Problems: Tests[6] were carried out in the full scale facility to simulate the station slurry transfer operations to determine the cause(s) of the slurry transfer/resin plugging problems at Bruce NGS 'A' mentioned earlier. It was determined that the problems were probably caused by a combination of slurrying procedures and 'passing' isolation valves in the inlet lines to the ion exchange columns. It was recommended that the inlet well screens (Johnson well screens with 0.008" slot width) be replaced with screens with larger slots to obviate the problem.

3. Resin Sampler: An in-line resin slurry sampler was developed[7] at the Research Division to collect a resin sample from a flowing resin slurry line that is representative of the contents of the column being emptied. The resin sampler was successfully tested at NPD (Nuclear Power Development) GS and Douglas Point NGS. We are undertaking a systematic program to collect and characterize spent resin samples from our nuclear stations. The program should help expand the database on spent resin characteristics, necessary for decision-making on ion exchange resin waste management strategies.

4. Volume Reduction Processes: Some research effort has been expended in the past on selected volume reduction processes for spent ion exchange resin wastes from CANDU reactor purification systems. An acid stripping process was developed[8] primarily for removal of carbon-14 from spent ion exchange resins, to be used as a pre-treatment step prior to a final volume reduction process such as incineration. The technique involves contacting spent ion exchange resins containing carbon-14 (eg, CANDU moderator resins) with hydrochloric acid to liberate $^{14}CO_2$ and absorption of the gas in alkali. The acid stripping technique has also been used for determination of C-14 concentrations on spent resins[9]. An electrically driven process, termed ElectroDialytic Resin Decontamination (EDRD)has also been developed[10] for volume reduction of spent PHT resins from CANDU reactors. The process involves removing radioactivity (primarily radiocesium) by passage of dc electric current through the spent resins in a specially designed electrodialytic cell. Pilot scale development of the volume reduction processes is planned for the coming years.

5. Alternative Sorbents: A program is just underway to evaluate various types of sorbents -- chelating resins, inorganic ion exchangers, crownethers, etc, to replace or augment conventional organic ion exchange resins in the nuclear purification circuits of CANDU reactors. The objective is to identify sorbents with very high selectivities for specific radionuclides that are of concern in the different purification systems in order to minimize replacement resin usage and hence minimize radioactive waste volumes produced.

SUMMARY

Ion exchange purification systems play an important role in CANDU reactor performance. Although the purification systems have operated satisfactorily over the years, there is scope for significant improvement in ion exchange performance in the systems. Research programs are underway to improve the performance of ion exchange materials and equipment in the various purification systems.

ACKNOWLEDGEMENT

The authors would like to thank D. Dawson of Ontario Hydro's Central Nuclear Services, for helpful discussions and suggestions in preparation of this paper.

REFERENCES

1. J.E. Lesurf, "The Water Chemistry of CANDU PHW Reactors". Paper presented at the Internal Conference on Water Chemistry of Nuclear Reactor Systems, Bournemouth, UK. October, 1977.

2. D. Barber and D.H. Lister, "The Chemistry of the Water Circuits of CANDU Reactors". Paper presented at the International Symposium on Water Chemistry and Corrosion Problems of Nuclear Reactor Systems and Components, Vienna, Austria. November, 1982.

3. R. Zimmerman, Ontario Hydro, Design and Development Division, Report No 83321, July, 1983.

4. T.J. Carter, "Radioactive Waste Management Practices at a Large Canadian Electrical Utility", Radioactive Waste Management, Volume 2(4), June, 1982.

5. B.R. Nott, Ontario Hydro Research Report No 84-136-K.

6. B.R. Nott, Ontario Hydro Research Report No 83-133-H.

7. B.R. Nott, Ontario Hydro Research Report No 83-147-K.

8. B.R. Nott, Ontario Hydro Research Report No 79-495-K.

9. B.R. Nott, "Determination of Carbon-14 in Spent Ion Exchange Resins", Int. J. Appl. Radiot. Isot. Vol 33, pp 584-585, 1982.

10. B.R. Nott, "Electrodialytic Decontamination of Spent Ion Exchange Resins from CANDU Primary Heat Transport Purification Circuits", paper presented at the International Symposium on Water Chemistry and Corrosion Problems of Nuclear Reactor Systems and Components, Vienna, Austria, November, 1982.

Ion-Exchange Experience at the Winfrith SGHWR

G.C.W. Comley and D.J. Ferrett, Senior Chemist, Winfrith Reactor and Head of Chemistry Division

INTRODUCTION

Modern power plants using steam as an energy transfer medium must be fed with make-up water of extremely high quality. This puts continuous and demanding requirements on industrial water treatment plants which can only be achieved by ion-exchange processes. The reliability, safety and economic operation of power plants especially those using nuclear boilers, is therefore directly dependent on ion-exchange materials. As well as being indispensable for the general production of large quantities of high purity make-up water, IX resins have found more specific uses in nuclear plants. Standard bead resins are not always adequate for the demands of such applications so that nuclear grade materials and powdered resins have been developed with the accent upon reproducible high quality.

The 100 MWe Winfrith reactor is a direct cycle pressure-tube BWR which uses heavy water as a moderator. The chemistry of both light water and heavy water is vital to the reactors effective and long-term operation. This paper describes the applications of ion-exchange resins in various systems of the reactor and highlights the more significant experiences over the 17 years of its operation. FIGURE 1 illustrates diagrammatically eight areas of application of ion-exchange materials in the Winfrith reactor. Consideration of the more important applications follows.

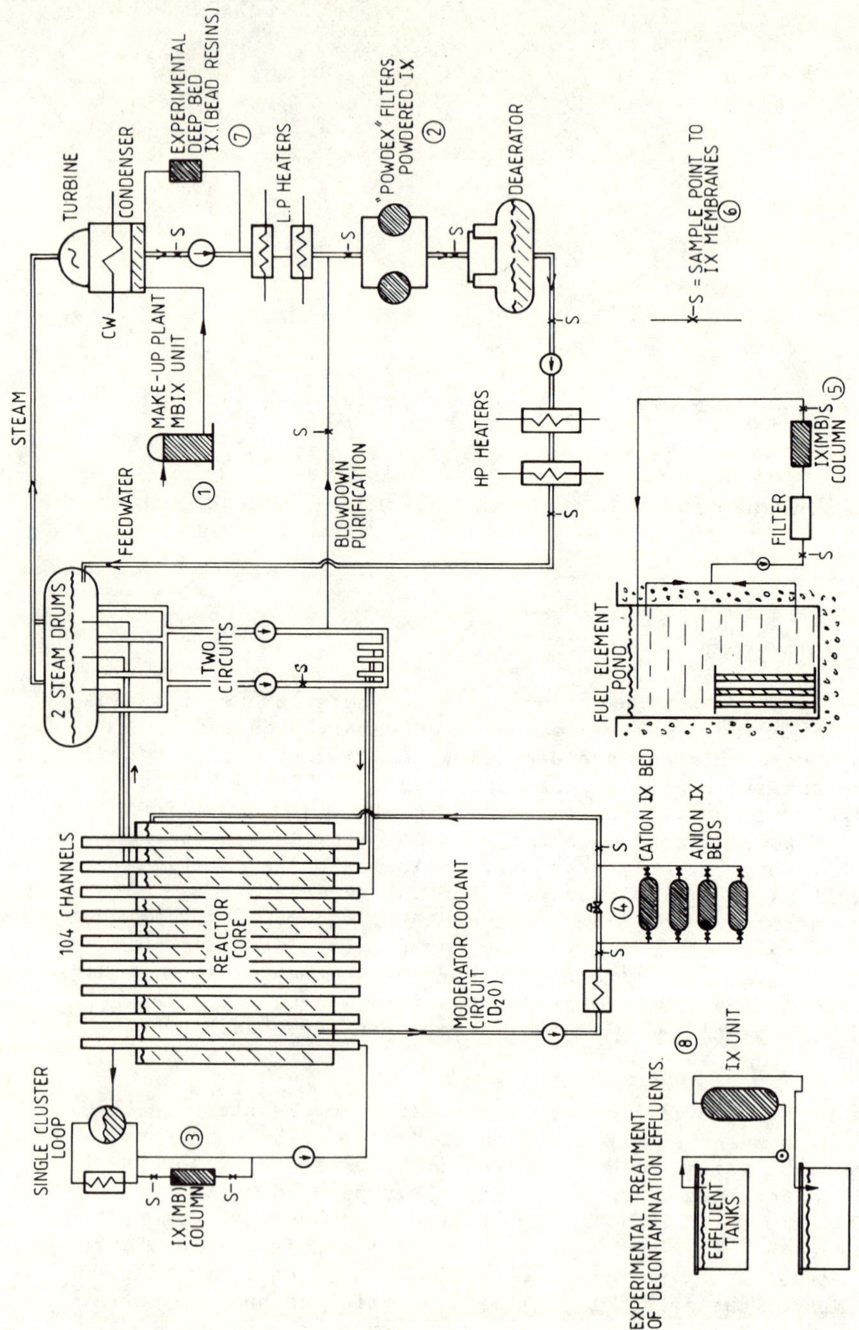

FIGURE 1 Applications of Ion-Exchange techniques at WSGHWR

FEEDWATER AND COOLANT PURIFICATION

When the Winfrith SGHWR was being designed in the early 1960's very little BWR experience was available worldwide. It was the operation of some first generation direct cycle water cooled reactors including WSGHWR itself, during the mid and late 1960's, which showed the need for better water purity standards than existed in most conventional power stations. Levels of corrosion and metal release which were acceptable in normal stations were found to be responsible for production of activation nuclides which led to unacceptable radiation levels in nuclear boilers. At the time however lack of water cooled reactor experience led the UKAEA designers to produce an elegant and costsaving way of treating the turbine condensate and coolant for the Prototype SGHWR. The reactor was provided with one filtration/ion-exchange plant to do both jobs, a unique feature as far as we are aware. Powdered IX resin had been introduced in the early 1960's in the USA as an improved system for steam condensate purification (1). The "POWDEX" process uses finely divided cation (H^+ form) and anion (OH' form) ion-exchange resin particles which are brought together in a recoat tank as a slurry. The resins then form a floc which is applied to the outer surface of filter leaves to form a coating of about 5 mm thick by recirculating the slurry. After the coat has been formed and the recoat tank water has cleared the filter can be operated to ion-exchange exhaustion or to a limiting pressure drop. Pressure drop across the unit increases as the coat removes particulate corrosion products. The spent resins in this process cannot be separated for regeneration and are discarded to waste.

On the basis of accumulating evidence (2) of the advantages of "POWDEX" resins it was decided to clean up the WSGHWR feedwater in mid-feedtrain. The plant, consisting of two filters each with 20 vertical stainless steel mesh leaves, would not only treat 330 t h^{-1} of feedwater but also about 22 t h^{-1} of coolant water. FIGURE 2 shows the location of the plant in relation to the feedwater and coolant circuits.

In summary the plant was designed for the following functions:-

(a) To provide minimum levels of soluble and insoluble impurities in feedwater and reactor coolant; the objectives being low corrosion of system materials and minimum fuel surface fouling.

(b) To operate at 105-110°C treating the entire feedwater flow except during recoat periods.

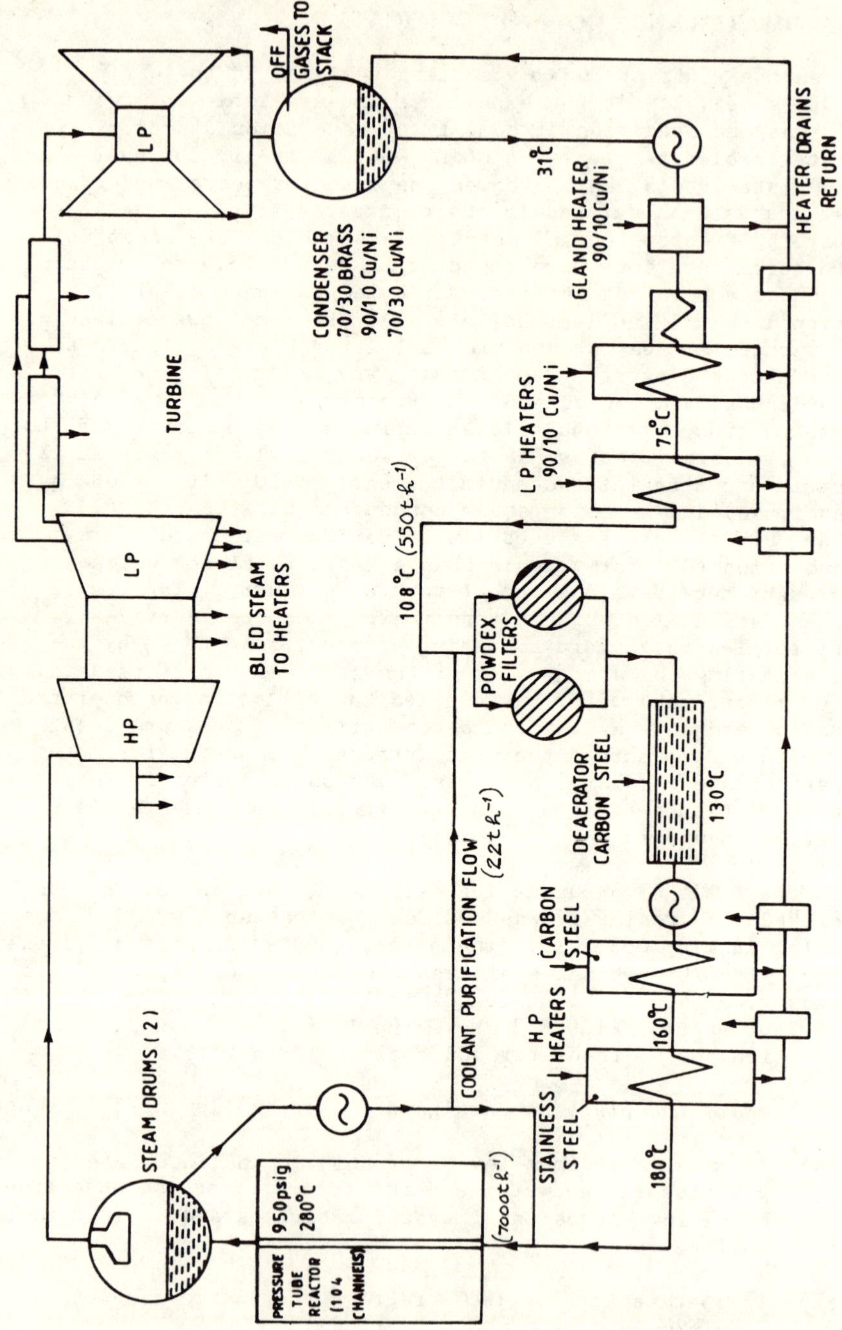

FIGURE 2 Simplified feedwater/coolant cycle for WSGHWR

(c) To allow for a limited variation in filter loading and anion to cation resin mix ratio and to operate up to a Δp of 1.2×10^3 N m^{-2} (25 psig).

(d) To be recoated at power. The "used" powdered IX resins to be non-regenerable and discharged to wet/solids active storage tanks.

After 17 years of operation the "POWDEX" purification plant has seen a diversity of non-standard conditions which have enabled its strengths and limitations (in terms of the Winfrith reactor design) to be recognised and documented (3). After some early modifications, the operators have learned to live with the plant as provided but it was clear that had the Commercial SGHWR been built a dual-function type plant would not have been repeated.

For this paper the operational highlights are referred to below. Additional details, including plant design aspects are given in Reference 3.

PERFORMANCE

Efficient chemical kinetics result from the large surface area and short diffusion paths associated with powdered IX resins (4). Filtration is also very effective because of the floc formed by the charged resin particles during the recoating procedure. The Winfrith reactor plant has consistently given outlet ionic purity levels approaching that of pure water, from 0.06 to 0.09 μS cm^{-1} at 25°C. This has enabled the coolant to be maintained at values below 0.5 μS cm^{-1}. More recently (since new filter leaves were installed in 1980-81) an average coolant conductivity of 0.2 μS cm^{-1} or less has been achieved over prolonged periods of reactor operation. This corresponds to a coolant chloride level below 20 mg t^{-1}, an order of magnitude lower than the operators specified limit for action to avoid stress corrosion cracking of stainless steel circuitry.

Iron species in feedwater and coolant are mostly insoluble Fe_2O_3 or hydrated Fe_2O_3, and some Fe_3O_4, the plant inlet concentration averaging 40 mg Fe t^{-1}. The removal efficiency of this particulate material (up to 20 μm) is 90 ± 5%. Copper is present primarily in soluble form at about 10 mg Cu t^{-1}. Of this, 80 ± 5% is removed by the Powdex plant.

Thermal degradation of the strongly basic amine groups present in powdered anion resin is more rapid than in bead-form resins (5). This might be expected when the relative particle sizes of the two types of resins are compared (FIGURE 3). Since the

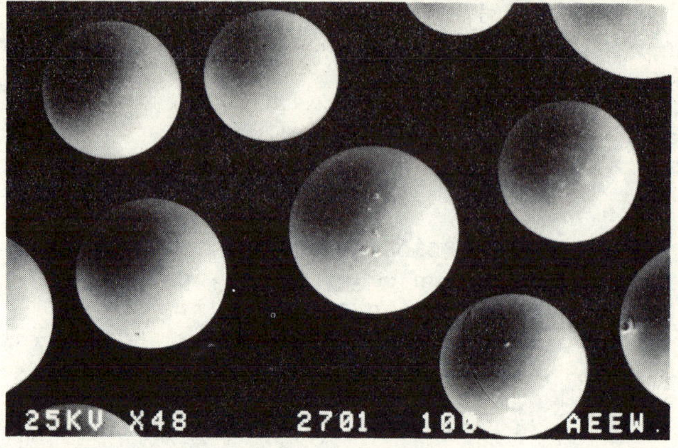

TYPICAL BEAD-TYPE RESIN (DUOLITE C255 H)

DIAMETER RANGE 450–650 μm

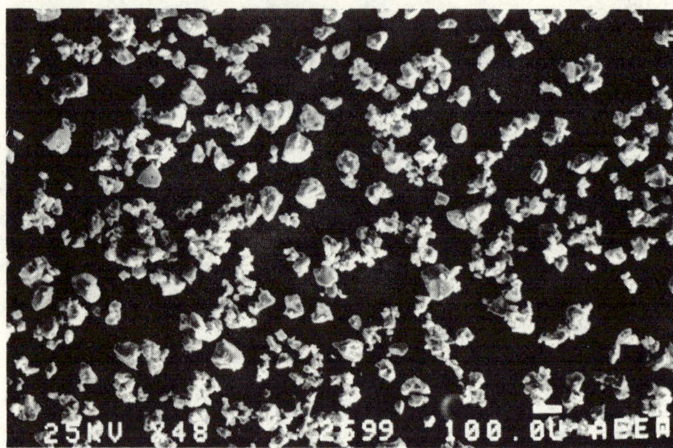

TYPICAL PREMIXED POWDERED RESINS (MICROIONEX MB 400)

PARTICLE SIZE RANGE 5–110 μm

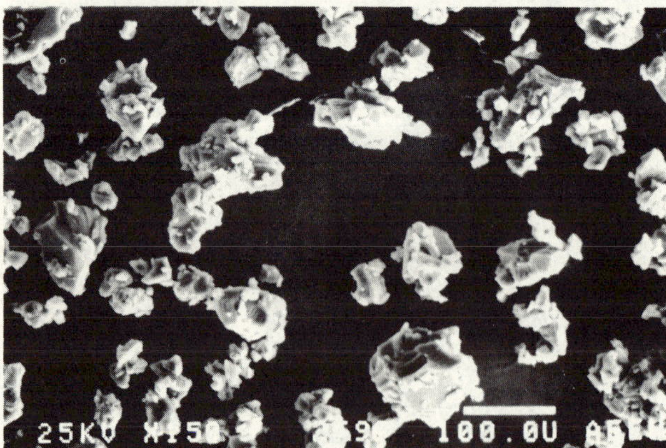

POWDEX CATION RESIN (PCH)

PARTICLE SIZE DISTRIBUTION:

<10 μm ~ 15% WT
10–20 μm ~ 35% WT
20–105 μm ~ 50% WT

FIGURE 3 Comparative sizes of bead and powdered IX resins

filters operate above 100°C the bed capacity for anions therefore reduces with time. Chloride is effectively retained between recoats (~ 2 to 3 weeks). The weaker soluble silicate ion is displaced after only a few hours at ~ 105°C. Because of the minimal capacity, silica is a major impurity in the WSGHWR coolant although no significant operating effect has arisen because of this.

Plant capacity for anions can be increased by lowering the feedwater temperature by valving out LP feedwater heaters at power or by increasing the anion to cation resin ratio in the recoat mix. Temperature reduction was used during a period of a small condenser leak in 1972. The normal resin mix is 2:1 cation to anion by weight. Large changes are not possible because of variations in floc properties which can adversely affect the rise of Δp in operation. At shutdown when the filters recirculate cooler water (~ 30°C) the anion resin capacity approaches accepted maximum values. This effect was useful earlier in the reactors' history for removing radioiodines present in the depressurised coolant (as iodide and iodate) after shutdown with failed fuel pins in core.

Corrosion products entering the reactor coolant circuit become activated and circulate as both soluble and insoluble species. Removal of activation nuclides (eg ^{51}Cr; ^{58}Co; ^{60}Co; ^{65}Zn; ^{59}Fe) occurs after the coolant blowdown stream (22 t h^{-1}) is diluted with feedwater (550 t h^{-1}). The efficiency of removal varies from ~ 65% for mainly soluble species to ~ 90% for those nuclides associated primarily with insoluble crud.

PROBLEMS

One early problem with the plant as commissioned resulted from poor distribution of the resin on the 38 m^2 of filter leaf mesh. In 1968, modifications to the internals of each filter unit provided a more even coating of resin and little by-passing of inlet water. A number of defects also developed due to fretting failure of the stainless steel mesh used to support the resin floc. Although this latter problem eased during the 1970's the failure rate of leaf mesh was never very acceptable and new leaf assemblies were designed at AEEW and installed in 1980-81 (FIGURE 4). Since that time only one minor leaf defect has occurred. Even minor tears in leaf mesh results in release of resin particles or agglomerates into the feedwater. These proceed to the coolant within a minute or so. The effects in the extremely pure and near neutral coolant are noticeable and immediate (FIGURE 5). They result from radiolytic breakdown of the active resin groups and to some extent of the polystyrene base material itself. The major effects occur as a newly coated filter is put on line. They may be summarised as follows:-

FIGURE 4 "Powdex" resin filter with new design leaves (1980)

FILTER RECOAT EFFECT	FILTER DEFECT	COOLANT LEVELS IRON $mg\,t^{-1}$	COPPER	OFF-GAS HYDROGEN $\%H_2$
MINOR Ⓐ	NO FAULTS	16 → 61	22 → 42	0.8 NO CHANGE
UNACCEPTABLE Ⓑ	SMALL LEAF TEAR	10 → 200	18 → 60	0.7 → 1.0
MAJOR Ⓒ	LARGE MESH TEAR	20 → 1250	20 → 160	0.7 → 1.1

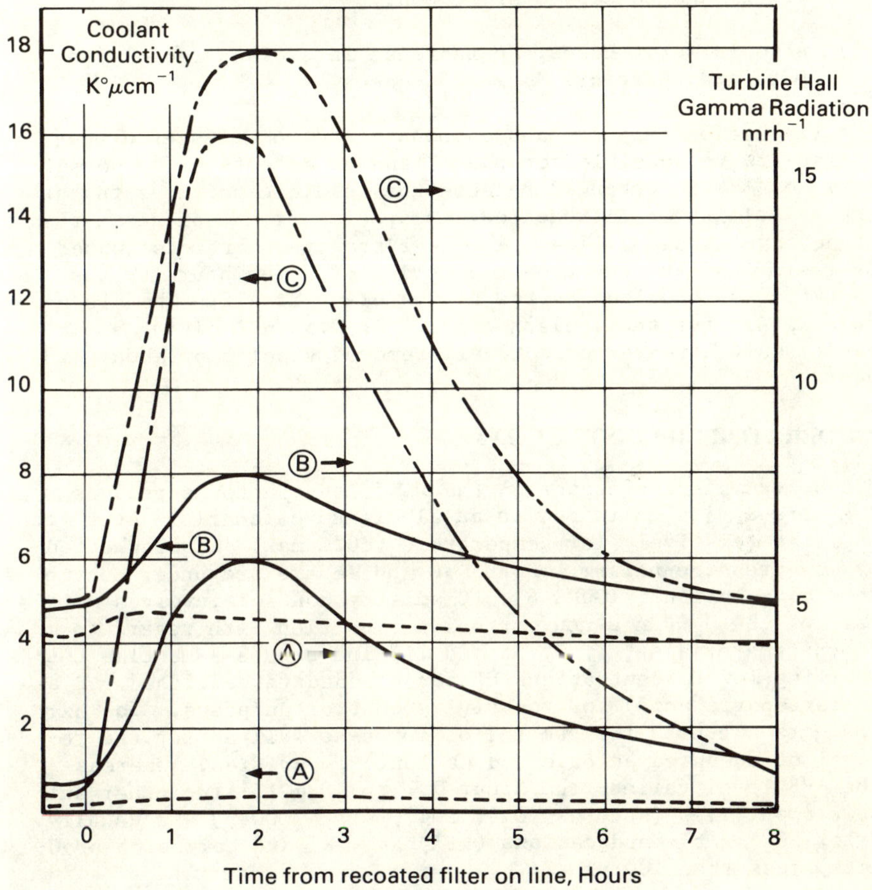

FIGURE 5 Powdex filter recoat effects – three levels of severity

(a) Increase in coolant conductivity and decrease in pH value.

(b) Increase in metal impurity levels circulating in the coolant (crud-burst effect).

(c) Increase in activation product levels carried by coolant.

(d) Increase in gamma fields around the turbine due to higher levels of ^{16}N carried by steam.

(e) In the worst cases, increase in steam condensate conductivity due to carryover of CO_2.

(f) Also in worst cases, an increase in coolant radiolysis rate and therefore H_2 in off-gases.

The release of sulphate and carbonate and to a lesser extent nitrate is responsible for the K° and pH effects. The normal level of ^{16}N is enhanced because the acidic conditions favours the volatility of nitrogen/oxygen species from coolant to the direct cycle steam. These chemical transient effects caused by resin ingress reduce over a period of a few hours by the purification provided by the Powdex plant itself. The clean-up constant for the coolant circuit is 0.55 h^{-1} giving an impurity half life for effectively removed species of about 1.3 hours.

MODERATOR CHEMISTRY CONTROL

The heavy water moderator in the Winfrith reactor surrounds the core and is contained in an aluminium calandria. At power it is at relatively low temperature (80°C max) and blanketted by low pressure helium. Both D_2O and He are are under continuous circulation (FIGURE 6). Chemistry control involves maintaining the D_2O in a non-corrosive condition with regard to circuit materials (Al; St.Steel; Ti) and at the same time controlling low concentrations of boron added (as enriched ^{10}B as deutero-boric acid) for reactivity control purposes. Ion exchangers are used for removal of boron as well as for the removal of unwanted anionic and cationic impurities. Whereas the ^{10}B level varies from about 0.2 to 20 g t^{-1} the other anionic impurities (such as NO_3', $SO4''$, Cl', $\{COO\}''_2$) are usually below 500 mg t^{-1} and cations (Li^+, Na^+, K^+ and corrosion products) less than 100 mg t^{-1}.

There are five resin beds of 0.11 m^3 capacity which operate at $<$ 40°C. Each is capable of treating up to 0.35 t min^{-1} from the 38 t heavy water inventory. One bed is in the cation (D^+) form, one anion unit is kept in the ^{10}boron satura-

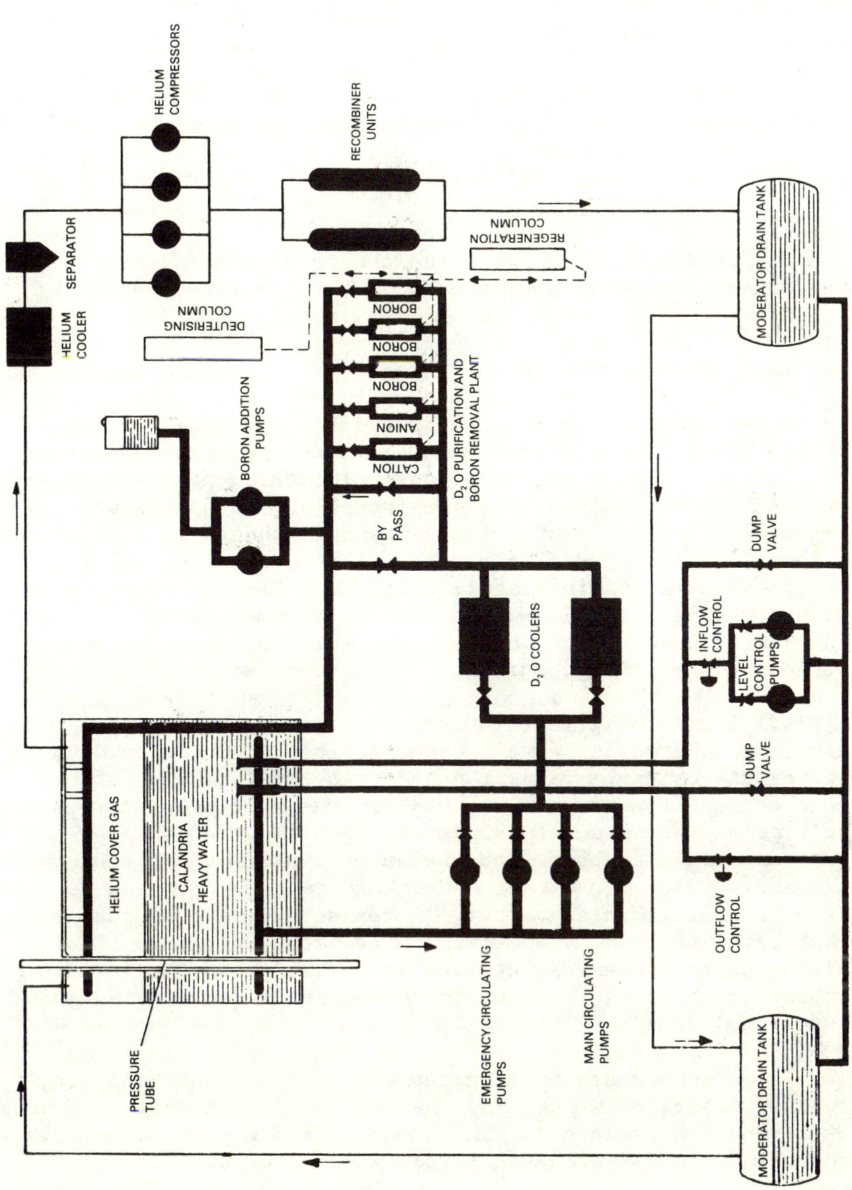

FIGURE 6 The WSGHWR moderator system

ted mode and three are available in the OD^- state for boron removal (FIGURE 5). Development work at AEEW in the years prior to reactor commissioning provided useful data on the capacity of strong anion resins for weak boric acid at low concentration levels. Operating experience has been quite satisfactory. For example a typical start-up after refuelling involves reducing ^{10}B from the shutdown level of 20 mg t^{-1} to about 8 mg t^{-1} with the calandria full of D_2O. It takes about 100 mins at the higher flowrate of ~ 0.32 m^3 min^{-1} to remove the necessary 450 gms of ^{10}B (FIGURE 7). Since each anion bed has a capacity of about 600 gms ^{10}B the need for 3 beds is evident. When saturated the boron removal exchangers are regenerated with KOD from which the ^{10}B is recovered by electrodialysis.

Impuritiy anions such as NO_3^- (from radiolytic fixation of N_2 in He) are removed as necessary by operating the saturated boron bed once or twice a week for a few hours at ~ 0.1 m^3 min^{-1}. It is necessary however to use the cation exchanger every day to control the level of lithium produced in flux from ^{10}B and to maintain an optimum pD at ~ 5.5. Experience has shown that the use of ion-exchangers should be kept to the absolute minimum and that high quality (nuclear grade) resins should be used to minimise impurity formation from breakdown of the resins themselves. For example, some batches of cation resins have proved to be susceptible to oxidation under prolonged operating conditions. The heavy water contains from 4 to 6 g t^{-1} of deutero-peroxide a product of radiolysis. It is possible to assess the extent of oxidative attack on the resin bed by measuring the reduction in D_2O_2 across the column. Oxidation leads to the formation of organic species (eg oxalate and CO_2) in the bulk moderator as well as the presence of sulphate. These impurities in turn put an additional burden on the anion control bed. Since the early 1970's, the cation exchange resin bed, which is used far more than any other, is replaced on a bi-annual basis, ie after about 550 EFPDs of reactor operation. This minimises radiochemical degradation effects and the level of impurities arising from the ion-exchanger base material and functional groups.

More detailed information on the chemistry of the Winfrith reactor moderator system and the role of ion-exchange resins is given in Reference 6. The chemical aspects of reactivity control are considered specifically in Reference 7.

C STANDARD USES OF IX AT WSGHWR

High quality demineralised water is used at the reactor primarily for coolant/steam/feedwater make-up purposes. Leakage

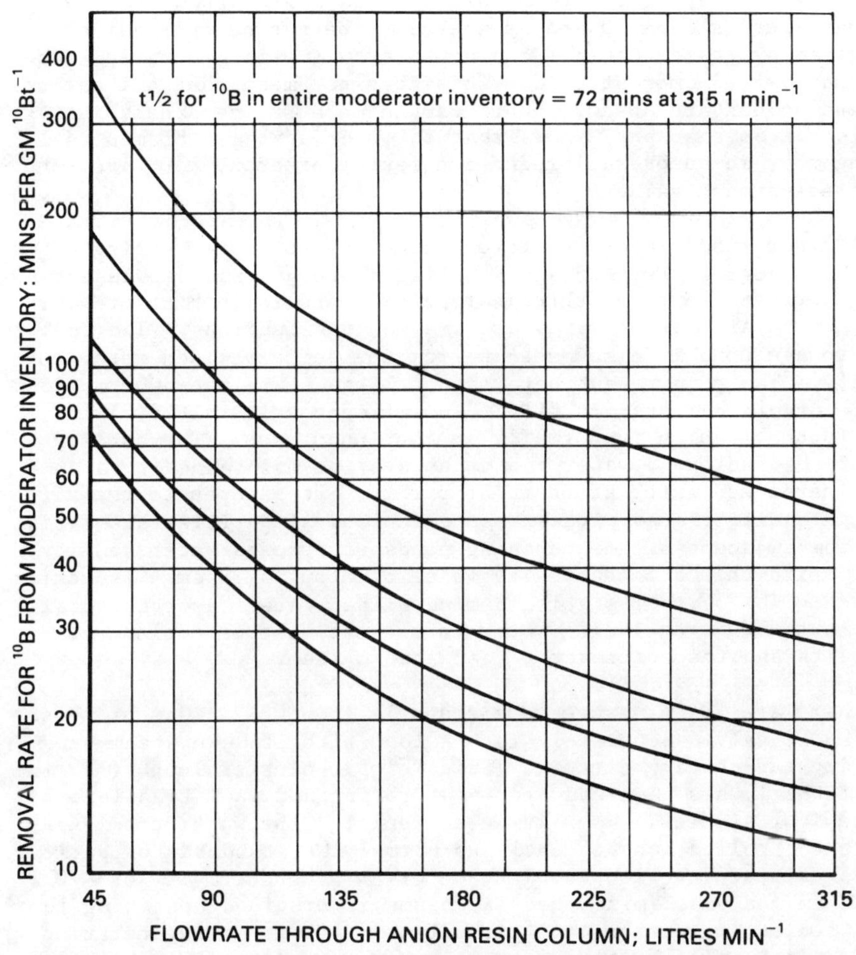

FIGURE 7 Removal rate for ^{10}Boron from Winfrith reactor moderator

from components in conventional steam raising plant has to be balanced by equivalent input of impurity-free water. The make-up plant shown at (1) in FIGURE 1 provides demineralised water for all station requirements at a maximum rate of 8 t h^{-1}. It should be noted that the out-reactor flow of steam and condensate is ~ 550 t h^{-1} at the full power of 100 MWe. For quality monitoring purposes this water must contain less than 20 and 50 mg t^{-1} respectively of chloride and silica from pretreatment levels of ~ 20 and 12 g t^{-1}. A better overall guide is given by its electrical conductivity which is generally less than 0.1 μS cm^{-1} at 25°C. This does not mean that all cations and anions are absent. For example more refined analysis (eg ion-chromatography) shows that chloride averages from 6 to 10 mg t^{-1}, ie an overall reduction factor of about 2500 over the pretreatment value.

Demineralised water is stored for emergency core cooling at the reactor. Two storage facilities are provided to inject impurity-free water through reactor channels in high pressure ($>$100 ATS) or low pressure emergency situations. The fuel storage pond is also made-up from the demineralised water supply. The requirement here is not large since the 3000 m^3 pond reservoir has its own filter and mixed-bed IX unit ((5) in FIGURE 1) which is normally in continuous use. The quality of the fuel pond water cannot however be maintained to a demineralised water standard of purity. It is open to the air and equilibriates with atmospheric CO_2. Impurities also arise from adjacent steam quenching ponds of process water quality. Despite this a conductivity value of about 5 μS cm^{-1} is achieved. The overall station demineralised water reserves total about 1300 t excluding water in the two reactor coolant circuits and the out-reactor feedwater system.

There are 104 pressure tube channels in WSGHWR. One of these can be valved separately into a loop with its own steam handling and cooling circuit. This "single cluster loop" ((3) in FIGURE 1) has been used for heat rating and fuel transient behaviour studies. When it is so isolated the water chemistry is controlled independently to normal BWR standards of purity. To maintain the loop coolant water at a conductivity of $<$0.2 μS cm^{-1} a small mixed-bed IX column is provided operating in a cooled leg of the loop circuit at ~ 40°C. No regeneration facilities are provided, the exhausted resins being discharged to the active wet solids drainage system for storage. Replacement resins are transferred to the IX column hydraulically.

Another application of ion-exchange materials relates to the analysis of various reactor water streams for trace metals and where applicable the associated radiochemical species ((6)

FIGURE 1). The technique of feeding a continuous stream of water from an isokinetic sampler through a capillary to a stack of membranes for filtration and cationic and anionic exchange was developed at AEEW for use on the Winfrith reactor in the early 1970's (8). It proved most useful in identifying and monitoring sources of cobalt being transported into the coolant circuits. This concentrating method also provides for integrated analyses for other trace metals such as Cu, Ni, Fe, Cr and Zn over the range of 0.005 to ~ 10 mg t^{-1} covering periods of continuous reactor operation of days to weeks.

DESIGN IMPLICATIONS OF REACTOR USES OF IX

Although the ion-exchange units at SGHWR are generally designed to conventional standards the need for shielding and remote control of valves etc is a feature of several plants. The powdered resin filters are housed in a shielded cell as part of the feedwater train since the presence of activation species in coolant and steam produces high background gamma fields. The ion-exchangers associated with D_2O purification, the fuel pond clean-up plant and the cluster loop are also sited in shielded cells in the reactor secondary containment. Those plants and resins associated with light water coolant and heavy water moderator are exposed to varying levels of gamma radiation from ^{16}N depending on flow rates. Whereas there are relatively low concentrations of activation nuclides in the moderator the input of corrosion products into light water coolant circuits produces nuclides such as ^{60}Co, ^{58}Co, ^{65}Zn, ^{54}Mn and ^{59}Fe which are removed by the Powdex plant. These build up between recoats as increasing sources of radiation such that the filter unit surface doserates can reach 10-20 Rh^{-1}.

DEVELOPMENT APPLICATIONS INVOLVING ION-EXCHANGE

When the Commercial SGHWR project was under consideration in the mid-1970's one aspect of the development programme involved the optimising of turbine condensate purification processes. In conventional and nuclear systems two types of full flow plant are used to counter condensate and feedwater impurities. These are based on bead resins (deep bed units) and powdered resins (filter demineralisers). There are also two main kinds of impurity to control, viz ionic species associated with cooling water inleakage and soluble/insoluble corrosion products released into the steam condensate system. Since the WSGHWR experience was limited to powdered resin plant it was decided to study the behaviour of the deep-bed system using an experimental bead-resin column ((7) FIGURE 1). While these studies were incomplete when the CSGHWR project was abandoned the following observations have since been confirmed in operational BWRs.

(a) Maximum initial protection against condenser leak impurities is provided by deep bed plants operating below 40°C.

(b) This reduces with time because of bead fouling effects due to insoluble (iron-based) corrosion products.

(c) Regeneration of the resins may not adequately remove insoluble metal oxides, physical methods may be needed.

(d) Resin separation, regeneration rinsing and remixing all have to be effective to prevent ionic throw from beds in subsequent operation at high flow rates.

and

(e) High flow rate deep-beds are not as effective as filter/demineralisers at removing insoluble corrosion products.

In more recent BWRs therefore both types of plant are often used in series for overall system protection and the purification of condensate. On the other hand some utilities chose to avoid the use of polishing systems lest they act as net generagenerators of the very impurities they are designed to remove (9).

Routine decontaminations of the Winfrith reactor coolant circuits has been practiced since 1969. Development work has involved the application of improved chemical reagents for removing crud from the fuelled circuits (10). Since 1980 the active corrosion products in LOMI (low oxidation metal ion) reagents has been standard practice (11). This reducing system consists of a dilute solution of vanadous formate and sodium picolinate. The latter acts as a buffer (pH 4.2 at 25°C) and was chosen specifically to keep iron and oxidised vanadium salts in chelated solution. The spent decontaminant therefore contains meallic cations, their activated species and organic acids. A series of ion-exchange experiments have been carried out on typical wastes, first in a laboratory rig and subsequently in a pilot plant scale experiment on bulk decontamination effluents ((8) FIGURE 1).

This work has demonstrated that effective removal of the main metals (V, Na, Fe) and major nuclides (^{60}Co, ^{65}Zn, ^{58}Co, ^{59}Fe, ^{54}Mn) is possible using strong cation resins at normal bed flow rates. Only a single pass is necessary to achieve $>$95% removal of these impurities within standard cation resin capacity values at ambient temperatures. Only chromium and metallic zinc species were less effectively removed in a single pass. The data obtained from this development work have been used in the preparation of a full scale plant design.

CONCLUSIONS

A wide range of applications of ion-exchangers is illustrated by Winfrith reactor experience. Although this is not fully comprehensive as far as all LWRs is concerned our overall experience may be regarded as typical. Integrity requirements for the safety of several water circuits as well as the need to minimise out-reactor doserates have led to very low specification limits for impurities that demand effective purification. Without ion-exchange it would be impossible to control reactor system chemistry and its effects on materials, radiation levels and reactivity.

It is essential however to understand the limitations of IX and in particular to allow for the sensitivity of organic based ion-exchangers used in the nuclear environment. Organic exchangers are complex chemicals in their own right and must be regarded as potential sources of impurity release when exposed to temperature, radiation or oxidising conditions. Here again WSGHWR experience has been mirrored at other nuclear stations in particular in direct cycle plants.

REFERENCES

1 J H Duff and J A Levendusky "Powdex - A New Approach in Condensate Purification". Proc Amer Power Conf 24, 739, (1962).

2 J A Levendusky and L Olejar "Condensate Purification Applications of the Powdex Process in High Pressure Utility Plant Cycles". Proc Amer, Power Conf 29, 840, (1967).

3 G C W Comley "Experience with Powdered Resin Purification at WSGHWR". Water Chemistry II, BNES, 1980, Paper 21.

4 F Martinola and G Kuhne "Properties and Application of Powdered IX Resins". Int. Conf on Ion Exchange in Process Industries, SCI, London, July 1969.

5 G C W Comley "Some Aspects of System Chemistry of BWRs". Prog Nucl Energy, Vol 2, No 3, 1978.

6 L A Chance "The Chemistry of the PSGHWR Heavy Water Moderator System". Water Chemistry I, BNES, London, 1978.

7 G C W Comley et al "Reactivity Control of the Winfrith Reactor". Nucl Energy, 1984, 23, No 3, June, 161-168.

8 H James "The Determination of Trace Amounts of Cobalt and other Metals in High Purity Water using Ion-Exchange Membranes". Analyst, April 1973, Vol 98, pp 274-288.

9 S D Strauss "Polishing Cuts Condensate Impurities below 0.1 ppb level". Power, October 1984.

10 G C W Comley "Coolant Circuit Contamination and Decontamination Experience at WSGHWR". ANS/CNA Meeting, Niagara Falls, Sept 1982.

11 D Bradbury et al "Decontamination of WSGHWR Coolant Circuits using LOMI Reagents". Nucl Energy, 1981, Vol 20, Oct, No 5, 403-408.

Ion Exchange Developments in Management of CEGB Fuel Ponds and Associated Wastes

D. Bradbury, Berkeley Nuclear Laboratories
D. Broadbent, North Western Region Scientific Services Dept.
G.A. Fitchett, North Eastern Region Scientific Services Dept.
T.J. Martin, North Western Region Scientific Services Dept.
D. Reed, Berkeley Nuclear Laboratories

INTRODUCTION

Ion exchange materials are used extensively in nuclear power plant for treatment of the fuel element storage ponds. The primary function of a cooling pond water treatment plant is to control the chemical conditions in the pond water to avoid or limit corrosion of the fuel element cladding [1]. It is almost inevitable that some radioactive species will be present in the pond water and therefore the treatment plant must also be capable of removing these species when necessary. A continuing programme of research has been undertaken to improve the performance of these plants which is the subject of this paper.

Besides the applications in pond water treatment another area in which ion exchange is potentially useful is the processing of effluents prior to disposal. A particular example is quoted in the paper in which radionuclides must be removed from a solution of 2 g dm^{-3} magnesium as bicarbonate solution resulting from the processing of Magnox fuel element debris.

MAGNOX POND CHEMISTRY AND TREATMENT PLANT

Spent fuel elements from the Magnox reactors are stored in cooling ponds containing a solution of 200 mg dm^{-3} sodium hydroxide. Water treatment is required to reverse the absorption of carbon dioxide into the pond, to maintain low levels of corrosive anions (Cl^-, $SO_4^=$ and $SiO_3^=$) and to control radioactivity levels. A water treatment plant on a Magnox cooling pond will in general consist of coolers, filters, non-regenerant caesium absorbers, cation exchangers, a degasser and an anion exchange bed. The ion exchangers are periodically regenerated giving rise to liquors which could be active in cationic or anionic species, e.g. Cs-137 and Sr-90 or Sb-125 and S-35 respectively.

AGR POND CHEMISTRY AND TREATMENT PLANT

Spent fuel elements on removal from Advanced Gas Reactor (AGR) cores are stored in cooling ponds containing a solution of 0.7% w/w boric acid and 100 mg dm^{-3} sodium hydroxide to allow short lived radioactive isotopes to decay prior to transport to the reprocessing plant. A chloride specification of 0.5 mg dm^{-3} has been set. During this storage period corrosion of the fuel can, dissolution of adherent oxide and leaching of the fuel element graphite sleeve may result in neutron activation and possibly fission products being released into the pond water.

The water treatment plant associated with AGR cooling ponds consists of coolers, filters and a mixed bed of non-regenerable ion exchange resins. The cation and anion exchange resins are soon converted to the sodium and borate forms respectively. Neutron activation species should also be removed from the pond water by the plant. If fission products are released into the pond then an alternative fission product specific ion exchanger must be employed.

CHEMICAL CONTROL OF MAGNOX AND AGR PONDS

Under the standard operating conditions of Magnox pond water treatment plant the influent pond water, containing sodium hydroxide and carbonate, is passed through a hydrogen form cation resin followed by a degasser to remove carbon dioxide. Aggressive anions are then removed by anion exchange and the water is then redosed with sodium hydroxide. However problems may arise when one or other type of active species predominates. On these occasions a modified water treatment plant operating regime may be necessary. Thus a restricted use of the anion exchanger (cation only) may reduce the effluent arisings of say S-35. Similarly a restricted cation unit regime (anion only) may be necessary to minimise Cs-137 and Sr-90 arisings. It was desirable to find out if the objectives of water treatment could be achieved by passage of the pond water directly through a hydroxide form anion exchanger, omitting the cation exchange, degassing and redosing steps.

The performance of an anion exchanger operating under the high pH conditions of this "anion only" regime has been investigated both in the laboratory and on the plant. The bed volume capacity for fresh and recycled ion exchange resin (IRN

78L) has been determined for SiO_2, $CO_3^{=}$, $SO_4^{=}$, Cl^- and K_H. The results showed that breakthrough of SiO_2 and Cl^- occurred together at 160 - 170 bed volumes. These were followed by carbonate breakthrough (200 BV) and sulphate (400 BV). The resin thus demonstrated adequate capacity for plant operation based on a daily regeneration cycle (160 g NaOH dm^{-3} resin). The "anion only" regime has been successfully applied for several months of operation at selected power stations with no loss of chemical control of the fuel element cooling pond. However a number of problems have been encountered and subsequently overcome. A gradual loss of chloride removal efficiency was observed after several weeks operation, e.g. 30% efficiency for an inlet concentration of 0.1 mg dm^{-3}. This efficiency was restored by converting the resin to the sulphate form and then regenerating with NaOH. This treatment is repeated about every six weeks. An unexpected problem was caused by the absorption of Cs-137 onto the anion resin. This resulted in about 15% of the inlet Cs-137 concentration appearing in the regeneration liquors each day. This was overcome by the inclusion of a non-regenerable guard bed (Lewatit DN) immediately upstream of the anion unit.

There was a need in AGR ponds to establish the ability of a borate form anion exchange resin to maintain low chloride levels and to remove chloride from the pond water efficiently following an accidental ingress. Experimental programmes have been undertaken where a number of parameters such as resin type, chloride concentration, and pH have been examined. At the recommended boron concentrations the dominating species appears to be the polymeric anion $B_3O_3(OH)_4^-$ and this is supported by the measurement of resin capacity for the boron containing anion. The capacity of the resin for chloride is therefore controlled by the relative affinities for this ion and the $B_3O_3(OH)_4^-$ ion on the ion exchange resin. The resin offering the best capacity appears to be a strongly basic anion exchange resin in a low chloride form, i.e. less than 0.5% of the exchange sites in the chloride form. IRN 78L is an example of such a resin. The results of the experimental programmes demonstrate that using this resin the chemical control of AGR pond water is feasible only for ambient chloride levels of the order to 0.5 mg dm^{-3} or less. In the event of an ingress of chloride such that pond concentrations exceed say 5 mg dm^{-3} then many bed changes will be necessary to return the chloride level to less than 2 mg dm^{-3}.

RADIOACTIVE CONTROL OF MAGNOX AND AGR PONDS

Control of radioactivity in Magnox ponds is achieved principally by the use, when necessary, of selective caesium removal resins in non regenerable units. The competition between caesium and sodium for the ion exchange sites is such that the resin must be chosen carefully to avoid excessive volumes of spent resin for storage and disposal. The most widely used material is an organic phenol-formaldehyde methylene sulphonic acid exchanger (Lewatit DN), though aluminosilicate materials have also been used [2].

A method of achieving an increase in the specific activity of discarded caesium selective resins has been developed. It involves absorption of caesium on a primary resin, elution of the activity with ammonium carbonate, evaporation and subsequent reabsorption on a secondary resin. Laboratory studies have been completed and a pilot plant constructed. The principles of operation have been fully described in an earlier paper [3].

A test programme has been devised to examine the bed life and efficiency of different ion exchangers for removal of neutron activation products from AGR ponds [4]. In the test programme particular emphasis has been placed on the removal of cobalt which represents $< 0.01\%$ of the total impurities by weight but accounts for ca. 20% of the total soluble pond activity. Four ion exchange types have been evaluated (Table 1), these represent the major types of synthetic ion exchange resins predicted to have a high selectivity for cobalt against a background of sodium.

After 6×10^4 bed volumes throughput the % cobalt removal for the four resins is shown in Table 2. As predicted the divalent cobalt was selectively absorbed by all the ion exchangers and it was necessary therefore to determine which resin had the greatest operating capacity by other means. This was achieved by scanning the columns containing the most suitable ion exchangers identified in the initial test (Zerolit 225, 236 and Duolite ARC 359) with a scintillation counter to determine the cobalt exchange profile. From these profiles it was possible to assess approximately the capacity of the ion exchangers and accurately determine their relative capacities. Zerolit 236, the weak acid ion exchanger, had a 30% greater capacity than any of the other candidate materials and its adoption would result in a significant reduction in storage requirements. A similar investigation has been undertaken to assess the performance of materials for fission product removal [5]. Caesium-134 and 137 have been identified as the likely major contributors for fission product activity in the AGR pond and it was predicted that the ion exchange materials having a high selectivity for divalent cobalt would be less effective for the removal of monovalent caesium. Natural and synthetic zeolites, and hexa-cyanoferrates are known to have a high selectivity for caesium and a number of these materials have been identified and evaluted in long term breakthrough tests employing synthetic pond water containing caesium-137

(Table 3). Column scanning has identified the natural zeolite AW500 as potentially the most suitable material for caesium removal.

TABLE 1 Ion Exchange Materials Evaluated for Cobalt-60 Removal

Zerolit 225	Strongly acidic cation exchange resin containing sulphonic acid groups on a 8% divinyl benzene crosslinked polystyrene matrix. Total capacity 4.8 meq/g dry resin.
Zerolit 236	Weakly acidic cation exchange resin comprising 2.5% crosslinked poly acrylic acid. Total capacity 10 meq/g dry resin.
Zerolit S1236	Chelating cation exchange resin with amino-diacetate groups on a crosslinked polystyrene matrix. Total capacity 4 meq/g dry resin.
Duolite ARC 359	Strongly acidic cation exchange resin comprising a phenol formaldehyde condensation product containing methylene sulphonic acid groups. Total capacity 3.5 meq/g dry resin.

TABLE 2 Cobalt-60 % Removal Efficiency After 6×10^4 Bed Volumes Operation

Ion Exchanger	% Removal Efficiency
Zerolit 225	100
Zerolit 236	99
Zerolit S1236	86
Duolite ARC359	100

TABLE 3 Ion Exchange Materials Evaluated for Caesium-137 Removal

Ion Exchanger	Structure	Total Capacity meq g^{-1}
AW500	Chabazite, effective pore diameter 4-5 Å	1.5
IE-96	Acid washed version of AW500	1.5
Zeolon 900	Mordenite, effective pore diameter 8-9 Å	1.3
Nickel Hexacyano Ferrate/Amberlite IRA904	Transition metal ferrocyanide precipitated in the pores of IRA904 anion exchange resin	~ 2.0

THE PROCESSING OF MAGNOX DISSOLUTION EFFLUENTS

One of the conditioning routes currently being assessed for the treatment of Magnox fuel element debris is the carbonate dissolution technique [6, 7]. The largest volume waste stream from this process is a solution consisting of a 2 g dm^{-3} solution of magnesium as magnesium bicarbonate plus a number of other impurities at significantly lower levels. In addition to these chemical impurities there are also amounts of various radionuclides present (e.g. Zr-95, Nb-95, Zn-65, Fe-55 and possibly Cs-137 and Co-60). The exact quantities of these radionuclides present depend to a large extent on the irradiation history and length of storage of the debris before conditioning. However, before discharge this effluent may be subject to chemical conditioning to reduce the level of radioactivity in solution. A number of ion-exchange materials have been assessed for this duty.

Experimental distribution coefficients and exchange break-through curves were acquired by batch equilibration and column techniques respectively, for a range of seventeen different materials of various categories. The results summarized in Table 4 refer to the seven top performers in the adsorption of radioelements uranium, cobalt, strontium and caesium.

TABLE 4 Relative Adsorption Efficiency Data for Principal Radioelements and Exchange Materials Selected by Batch Equilibration

Radioelement Adsorbed	Distribution Coefficient $K_d = \frac{\text{species concn g}^{-1}\text{ absorber}}{\text{species concn ml}^{-1}\text{ raffinate}}$						
	Inorganic			Organic			
	Manox A	Hydrous Titanium Oxide	Zeolon 900	Duolite A113 (Strong Anion Polystyrene)	Duolite A132 (Strong Anion Acrylic)	Duolite C433 (Weak Cation Acrylic)	Duolite ES346 (Amidoxime)
Uranium-238	Poor	Poor	Low	High	High	Poor	Poor
Cobalt-60	6.0×10^3	1.1×10^3	21	3.3×10^3	2.6×10^3	3.0×10^3	1.0×10^3
Strontium-85	8.8×10^3	7.4×10^3	2.1×10^2	< 1	6	2.0×10^2	7
Caesium-137	390	850	$> 5 \times 10^4$	10	18	19	6

The broad conclusions to emerge from our studies are:-

1) Hexavalent uranium (as UO_2^{++} ions) forms the anionic complex uranyl tricarbonate strongly adsorbed by the Duolite materials A113 and A132 in Table 4 (system capacity ~ 61 kg Um^{-3}).

2) Quantitative elution, with dilute HNO_3, of uranium from these organic resins indicates thermodynamic reversibility of the sorption/desorption cycle, which may be repeated without loss of adsorption capacity.

3) Cobalt can be removed by a number of materials, of which Manox-A (manganese dioxide) is the best. Column tests with ES346 have confirmed that this material is capable of removing Zn-65, Fe-59 and Co-60 from spiked dissolution solutions, particularly iron which displayed $< 5\%$ breakthrough at 1800 BV.

4) Strontium and caesium are best removed by Manox-A and Zeolon-900 respectively.

CONCLUSIONS

Ion exchange is widely used as means of chemical and radioactive control both in fuel ponds and in the processing of aqueous waste at nuclear power stations. This paper has described some of the research done to define optimum materials and operating regimes for ion exchange applications.

ACKNOWLEDGEMENT

This paper is published by permission of the Central Electricity Generating Board.

REFERENCES

[1] Windsor, B.W., 1982, Procs. of Conference on 'Gas Cooled Reactors Today', held Bristol September 1982, Vol. 1, p. 69.

[2] Broadbent, D., Haighton, A.P., Scanlon, F.B.E. and Smitton, C., 1979, CEGB Report NW/SSD/RR/105/79 (Unclassified).

[3] Bradbury, D. and Swan, T., 1981, Water Chemistry 2. British Nuclear Energy Society, London, pp 209-212.

[4] Fitchett, G.A. and Drury, T., Unpublished Work.

[5] Fitchett, G.A., Unpublished Work.

[6] Bradbury, D., 1983, Water Chemistry 3. British Nuclear Energy Society, London, pp 345-352.

[7] Bennett, D., Bradbury, D., Kaye, C.J. and Meese, R., 1984, Radioactive Waste Management. British Nuclear Energy Society, London, pp 69-79.

The Role of Ion Exchange in the Three Mile Island Cleanup

Richard M. Wallace, E.I. du Pont de Nemours & Co., Savannah River Laboratory, Aiken, SC 29808

INTRODUCTION

During the accident at Three Mile Island Unit 2 Power Station in March of 1979, a large amount of water accumulated in the sump of the Reactor Containment Building (CB) and overflowed into the basement. This water contained high concentrations of Cs-137, Cs-134, and Sr-90 caused by release from the reactor fuel, as well as high concentrations of sodium and borate ions. Water was also pumped from the CB sump into the Auxiliary and Fuel Handling Building (AB/FHB), flooding its basement as well.

The accident resulted in an accumulation of about 650,000 gallons of water in the CB and 550,000 gallons in the AF/FHB, while 90,000 gallons remained in the Reactor Cooling System (RCS)[1]. Typical values of the concentrations of the major contaminants in water from the CB and RCS are shown in

Table 1[2]. The composition of water in the AB/FHB was variable, but average values were about one-tenth that in the CB. The water in the three locations was decontaminated with two different ion exchange systems, the Epicor II System and the Submerged Demineralizer System (SDS), both of which will be described in this paper.

EPICOR II SYSTEM

The Epicor II ion exchange system was placed into operation in late 1979 to process the water in the AB/FHB so that the building could be decontaminated. The system consisted of three ion exhange columns (liners) in series assembled together with ancillary equipment in a shielded building that was originally designed for another purpose. The first two columns in the train were 4 x 4 ft liners each of which contained 30 cubic ft of mixed layers of inorganic ion exchange material and anionic and cationic organic ion exchange resins[3]. The third liner in the train was a 6 x 6 ft polishing column. The exact nature of the ion exchangers used in any of the liners has not been revealed because Epicor II was a proprietary system. The bulk of the radioactivity was adsorbed on the first of the liners while lesser amounts accumulated on the second and third.

THE SUBMERGED DEMINERALIZER SYSTEM (SDS)

During the summer of 1979, only a few months after the accident, the TMI-2 Technical Advisory Group selected a flowsheet for decontamination of the high-activity-level water in the CB.

This flowsheet was based primarily on the adsorption of the bulk radioactive contaminants, cesium and strontium, onto an inorganic ion exchanger. The exchanger was Linde Ionsiv™ IE-95, a chabazite-type zeolite that was commercially available and had a history of successful, large-scale usage. Zeolites were considered for this application because they have a much higher radiation stability than do organic resins; IE-95 was chosen specifically because it has a very high affinity for cesium and a fair affinity for strontium when in the sodium form. The flowsheet specified that the IE-95 was to be converted to the sodium form prior to use. Linde now markets the sodium form as IE-96. (Ionsiv IE-95 and IE-96 are trademarks of Union Carbide.) The early selection of the flowsheet was necessary so that process design and equipment fabrication could be started[4].

The processing system was designed by Allied General Nuclear Services for the Chem Nuclear Company, prime contractor for equipment fabrication and installation. The process was designed so that equipment containing high levels of activity was housed in one of the spent fuel handling pools in order to use the pool water for shielding. Therefore, the process was called the "Submerged Demineralizer System," or SDS, although the process was not intended to demineralize the water during its decontamination.

The zeolite and organic polishing resin were also ineffective in removing Sb-125 and Ru-106 which were present in anionic forms. These anionic contaminants were removed effectively, however, if the water was first deionized by removal of the sodium ions using a hydrogen-form cation exchange resin followed by adsorption of the contaminants on an anion resin.

The original SDS flowsheet called for the contaminated water to be clarified and passed through a series of four columns, the first three containing Linde Ionsiv IE-96, while a fourth contained an organic cation resin. A later revision had all columns containing the zeolite. The columns, which were modular, were intended to be used as waste containers after being loaded. They were to be moved after 12,000 gallons of water had been processed; this is equivalent to 200 bed volumes through one column. At that point, the column in the first position was to be discharged, each of the remaining columns moved up one position, and a new column installed in the last position. The limit of 200 bed volumes of water processed was dictated by the limited affinity of the IE-96 for strontium; considerations of cesium adsorption alone would permit over 1,000 bed volumes to be processed between column changes.

In 1981, a Department of Energy (DOE) Task Force[5] evaluated the technical and financial benefits of storing, shipping, treating, and disposing of the modular zeolite columns, assuming the columns were loaded with cesium to a higher level than called for by the reference process. The task force concluded that although it was technically feasible to load as much as 60,000 Ci of cesium on each column and to collect all of the cesium in the CB on 10 columns, the total number of columns could not be reduced to fewer than 27 because of the limited affinity of IE-96 for strontium.

Further work at ORNL showed that improved strontium decontamination could be achieved without affecting the cesium behavior too severely by using a mixture of Linde Ionsiv IE-96 and A-51, a zeolite with a high affinity for strontium. (A-51 is a trademark of Union Carbide.) Subsequent development showed that an optimum mixture for treatment of the CB water would contain 60% IE-96 and 40% A-51. That mixture was recommended for use in the SDS. A computer program was developed to describe column behavior during processing and to simulate the effect of column changeout. Small-scale column runs and computer calculations indicated that this mixture could process all of the CB water with only 10 columns and reduce both the cesium and strontium concentrations in the effluent to the level where contamination was controlled by residual colloidal species[6].

SYSTEM DESCRIPTION [1], [7]

The SDS water processing system consisted of two parallel trains of four 75-gallon vessels each containing approximately 60 gallons of zeolite. All ion exchange vessels were located approximately 25 feet underwater in the B-Spent Pool for shielding. Water was forced through the submerged filter vessels (sand beds) into four 15,000-gallon shielded feed tanks located in the A-Spent Fuel Pool. The water was pumped from these tanks through one of the two trains containing the zeolite-loaded vessels, through a post-filter to remove zeolite fines, and finally into one of two 12,000-gallon monitor tanks. These monitor tanks served as surge tanks from which water was transferred to the Epicor II system for polishing.

PROCESSING WITH THE SDS [1], [7]

The SDS was first tested with three 50,000-gallon batches of water that had accumulated on the Reactor Coolant Bleed Tanks (RCBT) as a result of decontaminating the AF/FHB. This water, which was contaminated to an intermediate level (<5 μCi/mL), was processed through Train 1 of the SDS. That train contained two vessels loaded with a 1:1 mixture of IE-96 and A-51. The average decontamination factor for both cesium and strontium was greater than 100. After leaving the SDS, the water was sent to the Epicor II System to further reduce the radionuclide content and thence to one of two 500,000-gallon Processed Water Storage Tanks (PWST).

During one of the successful manned entries into the CB, a submersible pump equipped with styrofoam flotation was dropped into the sump at an accessible location. The pump drew water from the surface and pumped it, using an existing penetration, through the filter vessels into the SDS feed tanks. Typically,

44,000-gallon batches were staged out of the CB to the feed tanks. The water was then processed through Train 2, containing four zeolite vessels loaded with a 3:2 mixture of IE-96 and A-51. The effluent flowed into a monitor tank from which it was transferred to the Epicor II System for polishing. After processing about 100,000 gallons of CB sump water, the modular zeolite vessels were moved countercurrent to the water flow as described previously. The method of vessel changeout was done to achieve maximum loading of cesium and yet remain below the shipping cask limit of 60,000 Ci. Samples of each bed influent and effluent were taken during processing of each batch to monitor bed loading and minimize breakthrough.

Processing of sump water began in late September 1981, with a small (15,000-gallon) batch that served to test the system with highly radioactive water. This was followed by a full batch, which loaded the first column to about 30,000 Ci of cesium. At that point, the vessels were changed out to test the vessel handling procedures. Subsequent vessels were loaded to levels closer to design values. Processing of CB water continued until early March 1982 when the surface suction pump lost suction with only one foot of water remaining in the CB basement. A jet pump was then installed in the basement and the rest of the water was pumped to the SDS feed tanks for later processing. In all, 650,000 gallons of water was processed with about 29,800 Ci of Cs-134, 278,000 Ci of Cs-137, and

11,600 Ci of Sr-90 activity removed. Most of this activity came to reside in eight vessels. The curie loading of the vessels involved in processing the CB are shown in Table 2.

During normal CB water processing, the effluent from the SDS was polished through the Epicor System and sent to the PWST for storage. The performance of the SDS and the Epicor System are shown in Table 3. The SDS was quite effective for removing Cs and Sr but ineffective for removing Sb, Ce, and Co. The Epicor System was effective for removing the residual activity except for H-3.

The contaminated water in the Reactor Coolant System (RCS) was processed through the SDS using two of the RCBTs for interim staging. Because the SDS removes only the cesium and strontium but passes the sodium and boron, it provides an ideal system

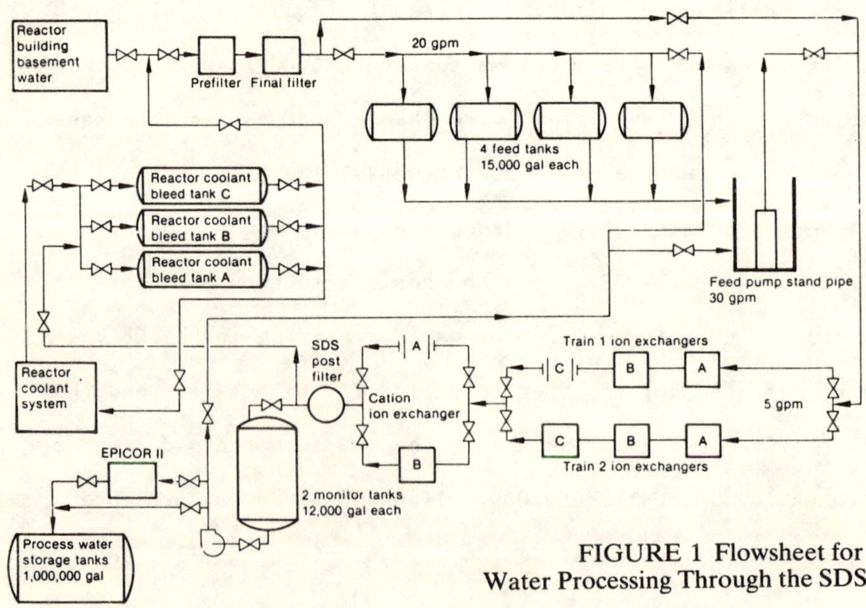

FIGURE 1 Flowsheet for Water Processing Through the SDS

for processing the RSC; the boron and sodium are necessary for criticality and corrosion control. The RCS was processed in a feed and bleed mode, because the reactor core had to be flooded at all times. The SDS and Epicor System were also effective in the cleanup of water used to decontaminate the reactor building.

TABLE 1 Composition of Contaminated Water [2]

Component	Reactor Coolant System	Containment Bldg. Water
Sodium	1350 mg/L	1200 mg/L
Boron	3870 mg/L	2000 mg/L
Nuclide	**Conc., μCi/L**	**Conc., μCi/mL***
H-3	0.17	1.0
Sr-89	5.0	0.53
Sr-90	25.0	2.3
Ru-106	0.1	0.001
Sb-125	0.01	0.02
Cs-134	10.0	26.0
Cs-137	47.0	160.0
Ce-144	0.03	0.0005

* Values corrected to July 1, 1980.

TABLE 2 Curie Loadings on SDS Liners After Processing the CB and RCBT[1]

Liner No.	Water Type*	Curie Loading Cs-(134,137)	Sr-90
D10015	RCBT	5,767	1,012
D10017	CB	30,312	1,021
D10012	CB	57,176	2,003
D10011	CB	44,317	2,061
D10013	CB	49,281	1,974
D20028	CB	43,333	1,660
D10016	CB	57,156	1,869
D20027	CB/RCS	4,289	5,0906

* CB = Containment Building; RCBT = Reactor Coolant Bleed Tanks
RCS = Reactor Coolant System

TABLE 3 Average SDS/Epicor Performance for CB Water Processing [7]

Radionuclide	μCi/mL Influent	SDS Effluent	Epicor Effluent	System DF
Cs-134	13	1.0E-4	<2.0E-7	>6.6E7
Cs-137	123	8.6E-4	3.2E-7	3.8E8
Sr-90	5.1	8.8E-3	1.7E-5	3.0E5
Sb-125	1.1E-2	1.1E-2	<4.0E-7	>2.7E4
Ce-144	4.0E-4	4.0E-4	<1.0E-6	>4.0E2
Co-60	2.0E-5	2.0E-5	<2.0E-7	>1.0E2
H-3	8.8E-1	8.8E-1	8.8E-1	0

ACKNOWLEDGEMENT

The information contained in this article was developed during the course of work under Contract No. DE-AC09-76SR00001 with the U.S. Department of Energy.

REFERENCES

1. H. F. Sanchez and G. J. Quinn. "Submerged Demineralizer System Processing of TMI-2 Accident Waste Water." U.S. Department of Energy Report, GEND-031 (February 1984).

2. D. O. Campbell, E. D. Collins, L. J. King, and J. B. Knauer. "Evaluation of the Submerged Demineralizer System Flowsheet for the Decontamination of High-Activity-Level Water at the Three Mile Island Unit 2 Nuclear Power Station." U.S. Department of Energy Report, ORNL/TM-7448 (July 1980).

3. P. Tuite. "Generation, Classification, Treatment and Disposal of Solid Waste Forms Resulting from Cleanup of TMI-2." In Three Mile Island Cleanup: Experience, Waste Disposal and Environmental Impact, L. J. King and J. H. Opelka, eds. AIChE Symposium Series 213, Vol. 78, 1982. pp. 22-26 (1982).

4. E. D. Collins, D. O. Campbell, L. J. King, and J. B. Knauer. "Water Decontamination Process Improvement Tests and Considerations." In Three Mile Island Cleanup: Experience, Waste Disposal and Environmental Impact, L. J. King and J. H. Opelka, eds. AIChE Symposium Series 213, Vol. 78, 1982. pp. 9-15 (1982).

5. DOE-SDS Task Force, R. E. Brooksbank, Chairman. Evaluation of Increased Cesium Loading on the Submerged Demineralizer System (SDS) Zeolite Beds." U.S. Department of Energy Report DOE/NE-0012 (May 1981).

6. D. O. Campbell, E. D. Collins, L. J. King, J. B. Knauer, and R. M. Wallace. "Evaluation of the Use of Zeolite Mixtures in the Submerged Demineralizer System (SDS) Flowsheet for Decontamination of High-Activity-Level Water at the Three Mile Island Nuclear Power Station, Unit 2." Paper presented at the international Zeolite Conference, Reno, NV (July 1983).

7. K. J. Hofstetter and C. G. Hitz. "Use of the Submerged Demineralizer System at Three Mile Island." Separation Science and Technology 18, 1747 (1983).

Index

Absorptiometry, x-ray, 89
Activated carbon, 37
Activated corrosion products, 171
ADAMO, 72
Adsorption efficiency data, 189
Air swept screens, 19
Americium, 124 125 126 133 138
AMEX process, 31

Baffles, 34
Bone-char, 119
BUFFLEX, 31
By-product-recovery, 36

CANDU reactors, 155 156 159 160 163
CHEMEX enrichment process, 48
 achievements, 49
 cascading, 50
 costs, 54
 economics, 55
 electrolyser, 52
 flowsheet, 50
 kinetics, 49
 plants, 53
 pulsed columns, 52
 technology, 51
 thermodynamics, 49
Chloride processes, 38
Citrates, complexing agents, 119 121
Coalescers, 35 37
Condensate purification, 179

Contactors, centrifugal, 37 71 80 90 96 98 99 130
Colete-Streat, 15
horizontal baffle, 91
mixer-settlers, 34 35 80 104 111
pulsed columns, 36 51 52 53 63 64 68
scale-up, 34
Cooling ponds, Magnox fuel, 14
AGR fuel, 15
Coordinating copolymers, 143 144
Copper, 146
Costs, capital, 16 21 35 36
operating, 23 24 25 33 35 36
Crud, 33 34 70
Crystallisation processes, 73

Deep bed ion exchanger units, 179
Degradation products, 129
Demineralised water, 177
Diluents, aliphatic hydrocarbons, 33
carbon tetrachloride, 126
Escaid 100, 33
flash point, 32 35
NPH, 126
perchloroethylene, 32
trichloroethylene, 126
Distribution coefficients, 149
americium, 139
neptunium, 139
plutonium, 139

EDTA, complexing agent, 120
Effluent purification, 180
Electrochemical extraction, 68
Electrolyser, 51
Electrolytic reduction, 63
ELKE, 68
ELUEX, 31
EMMA-2B, 71
Emulsions, 34 35
Entrainment, 33
Environment, 36 42
Extractants
alkylphosphates, 32
amines, 32 89
branch change amides, 89
DMPO, 124
cyanex, 272 36
DAPEX 3-1, 36
D2 EHPA, 36
DHDECMP, 121

DTPA, 121
HDEHP, 121
hydrolysis, 122 129
radiolysis, 69 90 117 129
TBP, 10 104 124
TOPO, 32 39

Feed clarification, 79 80 110
Feedwater purification, 167
Ferrites, preformed, 119
Filter leaf, 171
Fluid bed ion-exchamge, 19
Fuel pond purification, 178

Himsley column, 16
High level waste, 124 125
Hydrolysis, 129
Hydroxamic acids, 89
Hafnium, 71

Ion exchangers, 9 15 161
acidic salts, 135
anion resins, degradation, 169
applications, 165
capacities, 185
catin resins, degradation, 175
clinoptilolite, 118
columns, 158 159 193
Co removal, 186
Cs removal, 187
cyanoferrates, 135 136
fluid beds, 15
inorganic, 132 193
inorganic, natural, 133
oxides, 133 134
synthetic, 133
limitations, 181
membranes, 178
mixed bed resins, 158 160
organic, 132 186 193
performance, 137
preparation, 142
radiation stability, 141
resins, 15 17 18 32 117 120 156 157 159 161 162 163 167 184
sodium titinate, 120
systems, 155 156
titanium oxide, hydrous, 121
zeolites, 118 119 120 121 194
Ion exchange processes, 15 18
Iron, 32 110

Irradiated fuel bay (IFB), 155 156 160 161
Isotopic separation power, 53

Kinetics, 34
Kr cryogenic rectification, 72
Kr, R-12 wash, 72
KRETA, 72

LABEX, 68
Liquid metering, 81
Loop chemistry control, 178
Lysimeter, 37

Magnox fuel, 188
Mass transfer, 34
Mathematical models, 69
Metallurgical scrap waste, 124 125
MILLI, 68
MINKA, 69
Mixer-settlers (see Contactors)
Moderator, 155 156 157 158 159 160 162 163
    chemistry control, 174
Modifiers, 32
Molybdenum, 32 34
Neptunium, 139

PAMELA, 73
Packing, 34
pH control, 33
Phase separation, 32 33
Plutonium, 104 106 111 124 125 129 133 139
    electrochemical reduction, 68
    oxidation, 68
POWDEX, 167
    filter recoat effects, 174
    performance, 169
Process analysis, 70
Primary heat transport (PHT), 156 157 159 160 161 163
PUREX process, 67 90 124
    solvent, modified, 124 126 129 130
Purification systems, 155 156 158 160 163
PUTE, 68

Off-gas treatment, 65
Ores, processing, 31 38 40
Ozone formation, 72

Radionuclides, 38 42
Reactor coolant purification, 167
Reduction, 68

Remote maintenance, 21
operation, 96
Resin-in-pulp (RIP), 16 17 19 21 23 36
Rhodium, 145
ROXI cells, 68
Ruthenium, 145 150

Scrubbing, 33
Servo manipulator, 123
SGHWR, 165
Shear, whole bundle, 119
Shielding, ion-exchange plants, 179
Solvent-in-pulp (SIP), 35 40
Solvent losses, 33 35 36
recovery, 35
volatility, 36
Spectrometry, 89
Suspended solids, 15 18
Submerged demineralisation, 194
Stripping, 32 33 40
agents, 33 34 40

TBP (see Extractants)
TDR, liquid level measurement, 89
Third phase formation, 126
Three Mile Island, 192
TRAMEX, 12
TRU elements, 11 124 125 127 128 130
TRUEX process, 126 128 129 130

Uranium
enrichment, 11 48
ore, 11 15 16 17 18 31 119
production costs, 14 18
Uranium-plutonium separation, 68 104
UTE, 68
Vanadium, 32

WAK plant, 68
Water chemistry, 155 165

X-ray absorptiometry, 89

Zirconium, 32